CLASSIC
FARM TRACTORS

CLASSIC FARM TRACTORS

MICHAEL WILLIAMS

SPECIAL PHOTOGRAPHY

Andrew Morland & Terence J. Fowler

BLANDFORD PRESS
POOLE · DORSET

First published in the UK 1984 by Blandford
Press, Link House, West Street, Poole, Dorset,
BH15 1LL

British Library Cataloguing in Publication Data

Williams, Michael
   Classic farm tractors
   1. Farm tractors—History
   I. Title
   629.2'25   TL233

ISBN 0 7137 1421 2

Typeset by August Typesetting, Haydock,
St Helens.

Printed in Spain by Printer Industria Gráfica,
Barcelona D.L.B. 29312-1984

CONTENTS

INTRODUCTION

This is another selection of tractors chosen from more than 80 years of development.

Some of the classic tractors included here were commercially successful – part of the history of mechanisation on thousands of farms in many parts of the world. Some helped to introduce important new developments which have raised the efficiency of power farming and are now familiar features on almost every tractor.

Many more of these tractors were neither successful nor important. Some were simply the right idea at the wrong time and others are examples of the many odd and often eccentric ideas which have helped to make tractor history so fascinatingly varied.

I have included tractors powered by electricity and steam, fuel-cell power units and a gas turbine. There is a tractor equipped with closed-circuit television equipment and others designed to cultivate by means of a cable system. There is even a tractor which can cross the sea.

The successes and the failures are all part of tractor history and all of them represent individual ideas of how power farming should develop. Together with the publishers, I am grateful to Andrew Morland and Terence J. Fowler for their excellent photography and all concerned are indebted to Mr Les Blackmore of the Hunday National Tractor Museum in Northumberland and Mr Terry Farley of the Manitoba Agricultural Museum in Canada for their fullest cooperation in arranging for the tractors to be photographed.

I hope there is something in this collection of pictures and descriptions for everyone with an interest in tractor history.

Michael Williams

IVEL

Farm tractors made their first appearance in the USA, but it was a group of British designers who were the first to realise the full potential of the tractor as a really versatile, general-purpose power unit.

American ideas about tractor design were heavily influenced by the needs of the big-acreage grain farms. These were the farms where steam power had started to make a significant impact and American tractor manufacturers were in direct competition with the steam engine companies.

The situation in Britain was quite different. Farming economics, and even the field sizes and soil types, ruled out the heavyweight approach to mechanisation, and demanded more originality in tractor design.

Soon after the turn of the century, there were plenty of people in Britain with original ideas, who produced prototypes and, in some cases, production tractors of outstanding interest and imagination. Between them they introduced many of the features which have become essential characteristics of the modern tractor.

Between about 1902 and 1905, Ransomes built a tractor with three gears forward and three reverse, and with stub-axle steering; Scott tractors appeared with a power take-off and with a powered rotary harrow and seedbox; Sharp built a powered grass cutter; and Petter and Saunderson both built transport tractors with load-carrying platforms.

The greatest, and also the most successful, of this group was Dan Albone, who built his tractors at Biggleswade, Bedfordshire, and used

Ivel Tractor name-plate photographed on the Hunday Museum tractor.

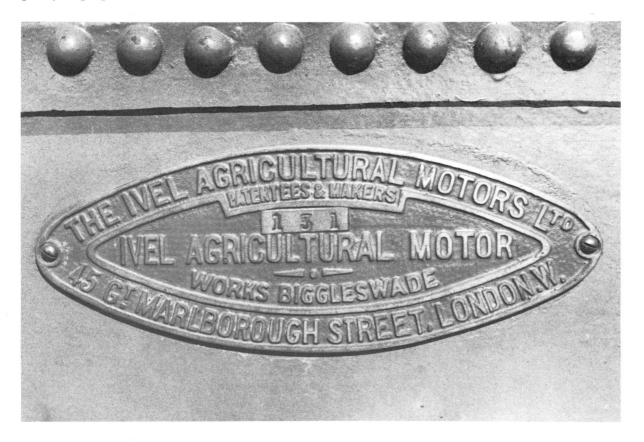

Driver's-eye view of the Ivel.

the trade name Ivel, which is the name of the river which runs through the town.

Albone was the son of a market gardener, but his interests and talents were in engineering. By 1880, he had started his own small business as a cycle maker and repairer. The business developed rapidly, helped by some design improvements which Albone patented and by his personal successes in cycle racing.

His engineering ideas developed beyond bicycles and he began building axle bearings for the Great Northern Railway, and built several cars and motor cycles.

There is evidence that he was interested in designing a tractor as early as 1896, and he had a prototype working in 1902. His first tractor had many of the design features which remained throughout the production life of the Ivel.

It was a three-wheel tractor, with a single front wheel (Plate 1). The twin-cylinder engine was mid-mounted, with a cooling tank beside the driver and over the rear axle. Early production models were described as 8 hp, but more powerful engines were fitted later.

One of the features of the tractor which brought favourable comment in Britain was its light weight. At about 25 cwt, it was considered that there was little risk of soil damage by compaction. The tractor also earned praise for the simplicity of its controls. In 1902, there was no such person in Britain as a skilled tractor driver and ease of operation was an important feature.

Dan Albone's skill as an engineer was matched by his ability as a salesman. British agriculture was not ready for the tractor and Albone and his

The Hunday Museum's Ivel, probably the oldest tractor in working order.

competitors had to work hard to find customers.

One idea to encourage interest in the Ivel was a series of demonstrations. These started in 1903 and were apparently held regularly twice each month, on a farm near Biggleswade where a range of equipment to be used with the tractors was kept permanently.

Albone also showed considerable skill in maintaining a flow of press publicity for his tractors. Export successes, which were numerous, were frequently reported and his tractors won an impressive list of medals and honours at shows throughout Britain and overseas; these also meant additional news items for the press. He

also earned extra publicity by providing performance figures and costings for his tractors. These were checked by independent witnesses and included a comparison between tractors running on petrol and alcohol.

He made ingenious efforts to find alternative markets for his tractors as the demand from British farmers remained so disappointing. One of his tractors was demonstrated as a fire engine and a bullet-proof model was shown to army officials in a bid to promote the tractor as a first-aid unit in wartime.

Dan Albone died in 1906, aged only 46. He was one of the most remarkable people involved in tractor development and he achieved more in just a few years than most tractor designers achieve in a lifetime.

BRUTSHKE ELECTRIC TRACTOR

While engineers in the USA and Britain were developing tractors with internal combustion engines, some people in Germany were becoming side-tracked into designing tractors powered by electricity.

This was confined mainly to the sugar beet-producing areas. Beet-processing factories sometimes had their own electricity-generating equipment and, for much of the year, this was under-used by the factories. The factories were apparently willing to sell some of this surplus 'off-peak' current and several companies built tractors and ploughing engines which could make use of the power supply.

Electric ploughing equipment was being used in north Germany in 1894 and possibly even earlier than this. It was still attracting enthusiastic interest in 1902 when the Brutshke system was introduced.

The Brutshke engine was supplied with current which was distributed from the nearest sugar beet factory through cables carried on tall wooden poles. The electric motor was mounted on a large four-wheeled chassis, which also carried a winding drum for the ploughing cable and a large quantity of electrical cable to link the tractor unit with the nearest main supply line.

A transformer was also carried on the tractor unit to reduce the voltage from the 2200 volts supplied from the factory to the 220 volts needed by the motor. The motor could power either the winding drum or the rear wheels of the tractor, but not both at the same time.

The Brutshke system differed from previous electric ploughing equipment because it required only one engine, with an anchor on the opposite headland. A two-way balance plough was pulled to and fro between the engine and the anchor, turning three furrows in each direction.

Double-engine ploughing sets, popular in Britain with steam power, had an engine and winding drum on each headland. A report on the Brutshke system said that the operating costs were similar to those for a double-engine set of equipment as the labour requirement was a

Balance plough used with the Brutshke electric cultivation system.

(Top) This was the anchor which held the cable in the Brutshke system.

(Above) Brutshke electric tractor and windlass.

three-man team for both. With only one cable engine, the man on the second engine had to be replaced by someone to see that the anchor moved forward for each new set of furrows. The third member of the team rode on the balance plough.

The capital cost of the Brutshke single-engine system was obviously much less than for a double-engine set, but there were disadvantages when only one engine was used. One problem was that the man on the single engine had complete control of the plough in both directions. In a small level field, the operator could see when the plough was approaching the anchor on the opposite headland and could stop it and reverse the winding drum at the right time.

On an undulating field, or in foggy weather, the operator might have difficulty in deciding when to stop the plough, and a mistake would send the plough and its ploughman crashing into the anchor.

A report in the British journal, *Implement and Machinery Review*, in 1903, was enthusiastic about the Brutshke system, which it described as ideal for flat fields.

'The economy and efficiency of this method of electrical cultivation have been abundantly demonstrated for some time past on beet fields adjacent to sugar and other factories, and it is a question even whether it would be advantageous on extensive farms to lay down an electric generating plant.'

Although the Brutshke engine was designed primarily for cable work, it could also be used as a tractor for pulling trailers and other equipment. But its operations were severely restricted by the trailing cable and the need to be near an electricity supply point.

These problems put the electric tractor out of business in Germany, just as they did in the USSR and other countries where similar systems were tried.

DRAKE AND FLETCHER

The Drake & Fletcher tractor seems to have had a brief commercial existence. It made several public appearances in 1903 and was described in the farming press of that year. Then, apparently, everyone lost interest in it.

It is likely that only one tractor was completed. It was built in the company's factory at Maidstone, Kent, where Drake and Fletcher still manufacture sprayers and sell a wide range of tractors and equipment.

The company, known as Drake and Muirhead until the arrival of Mr Fletcher as a partner in 1898, had been building oil engines since about 1885. The engine used in the tractor was a Daimler and was unusual in having three cylinders. Few details of the engine have survived, but it was described as being 'guaranteed remarkably free from vibration', with a rotary magneto ignition system which gave 'excellent results and instantaneous starting'. The output was 16 hp.

There were two forward ratios and a reverse, with top speeds of 3 mph in first gear and 6 mph in second. The final drive was through roller chains of 2 in pitch. To stop the tractor, there was a footbrake operating on the countershaft, with band brakes to the rear wheels, operated by hand.

Photographs show that the tractor was short and high. The overall length was 8 ft, and the height to the top of the engine cover was 5 ft 6 in.

The first public demonstration took place during the 1903 ploughing match organised by the North Kent Agricultural Association, claimed to be the biggest event of its kind in the world. On its demonstration plot, the tractor pulled a two-furrow plough at 3 mph. The tractor also demonstrated its usefulness by bringing a trailer from the factory to the ploughing match with a load of equipment for display on the company's stand.

A report in the *Hardware Trades Journal* said the tractor was 'designed for all kinds of farm and general estate work, including ploughing, cultivating, reaping, binding, mowing, hop washing, etc. and is also capable of being used for stationary work'.

The price for such a versatile tractor was £375, or £350 according to another magazine report.

(Above) Drake & Fletcher tractor; the radiator
was at the rear.

(Right) Front view of the Drake & Fletcher
showing the three-cylinder engine.

HART–PARR 'OLD RELIABLE'

Hart–Parr's 'Old Reliable' (Plate 2) is one of the best known of the heavyweight tractors which helped to mechanise big acreage farming in Canada and the USA. Tractors of this type were a spectacular phase in the history of power farming, and the surviving examples are among the most popular attractions in the big agricultural museums.

They were also expensive, temperamental and primitive, according to their critics, and there is evidence to support this view. The authors of a research paper published in 1980, *The Adoption of the Gasoline Tractor in Western Canada*, suggest that many of the farmers 'who had the courage or the foolhardiness' to buy one of the early tractors, regretted the investment.

Instead of the extra profits suggested by the salesman, the tractor owner faced reliability problems, poor service back-up from the manufacturer, high operating costs and difficulties on wet soils as the heavy machine easily bogged itself down.

Hart–Parr specialised in building heavyweights from their first tractor in 1901 until market conditions forced them to introduce smaller models after the war. The 'Old Reliable', built from 1907 to 1918, was powered by a 60 hp engine and weighed 10 tons. By Hart–Parr standards, it was a medium-size model and their biggest tractor was the 60–100 which weighed 26 tons.

The engine which powered the 'Old Reliable' was a two-cylinder, horizontal design, starting on petrol and running on paraffin. It developed its rated power at a sedate 300 rpm from cylinders with a 10 in bore and 15 in stroke. Five dry cells provided the spark for starting the engine, with a low tension magneto to keep it running.

In the Hart–Parr instruction book, there was a nineteen stage sequence of operations for starting the engine from cold. These included turning the 1000 lb flywheel by hand to spin the engine. Even when the full procedure was correctly followed, engines of this type were not always easy to start. This was partly because of the unreliable ignition and carburation systems, and partly because of

Hart–Parr 'Old Reliable'.

inadequate standards of maintenance on many farms where the tractor driver had only recently been promoted from driving a team of horses.

Jokes about the farmer who left his tractor engine running all night, rather than face the complications and uncertainties of starting it again next morning, were probably based on at least a grain of truth.

The primitive engineering standards are apparent from the photograph, which was first published in 1911. The chain-operated steering mechanism, inherited without modification from nineteenth-century steam-engine design, was hard to use and lacked accuracy. There was plenty of exposed gearing to collect mud, dust and small stones, and the driver's forward vision was largely obscured by the massive rectangular cooling tower over the front axle.

The gearbox offered just one ratio forward and one for reverse, with a maximum road speed of 2.3 mph. In later models, the travel speeds were raised to 4 mph or even more.

In spite of the criticisms, the 'Old Reliable' was one of the most successful tractors of its type and apparently one of the best liked. It remained in production throughout World War I and played a significant part in bringing prairie land into cultivation for the first time.

MARSHALL 30 AND 60 HP

Throughout most of its history, Britain's tractor industry has relied heavily on export business. This was particularly true of the early years of tractor development in Britain, when the home market provided few customers, and manufacturers such as Ivel and Saunderson built up overseas markets.

Marshall was another tractor company which had realised the importance of export business at an early stage. The first Marshall oil tractor was announced in 1907 and, by that time, the company was already well established in countries such as Australia and Canada as an exporter of steam traction and portable engines.

The 1907 tractor was described as a 30 hp model. The power unit was a two-cylinder design with individually water-jacketed, vertical cylinders of 6.5 in bore and 7 in stroke. The engine was designed to operate at 700 to 800 rpm and ran on paraffin.

The gearbox provided three forward ratios, with speeds of 2, 4 and 6 mph, with a gear drive to the rear wheels. A large tank of cooling water and the fuel tank were both located so as to provide extra weight over the rear axle, and there were seats for the driver and the passenger over the water tank.

Cooling the engine appears to have been a complicated business. The water circulated from the water jackets to the top of the radiator, where the engine exhaust was used to induce air movement. A pump moved the water from the base of the radiator back to the rear tank, ready to circulate forwards to the engine again.

According to the manufacturers, the cooling system used between 2 and 3 gallons of water a day, due to evaporation and 'inevitable leakage', and there was enough capacity in the system to allow for this without topping up during a full day's work. This claim was not borne out when the Marshall took part in the 1908 competition at Winnipeg, where a slightly different version of the tractor used 18 gallons in a 2 hour hauling test. This was not an excessive amount by comparison with up to 32 gallons used by some of its competitors.

The Winnipeg appearance was the first of a determined sales campaign which continued in Canada until after the beginning of the war in 1914. The campaign appears to have been reasonably successful against the very large number of American manufacturers which were also becoming well established in Canada. One result of Marshall's efforts is that some Marshall tractors have survived in Canada, and most of the big agricultural museums have one while none is

An early version of the 30-hp Marshall.

known to have survived in Britain.

At the 1908 Winnipeg event, the Marshall entry took third place in an entry of seven tractors. This apparently encouraged the company to make a more ambitious effort the following year, with three entries. These were the 30 hp tractor, a steam traction engine, and the new 60 hp oil tractor which was to become their biggest selling model in Canada.

Results for the 1909 event were less favourable for Marshall. The steam entry had to be withdrawn, the 30 hp tractor was placed third out of three entries, and the 60 hp oil tractor came second in an entry of five. The 60 hp tractor did much better at the newly established tractor competition at Brandon, Manitoba, in the same year. In his book, *Power for Prairie Plows*, Grant MacEwan quotes an advertisement for 1909 in a farming magazine, in which Marshall commented that they had 'been awarded the gold medal for their 4 cylinder engine at the Brandon Fair which, as all farmers know, is a much more important machinery exhibit than Winnipeg'.

The 60 hp Marshall weighed almost 10 tons, which was twice the weight of the 30 hp model. The four-cylinder engine was a pair of the two-cylinder power units placed side by side. The selling price in Canada, inflated by the high cost of shipping the tractors from England, was $3400, compared with $3250 for their 60 hp

steam traction engine and $2500 for the 30 hp tractor.

When the 60 hp tractors were sold in Canada, they were equipped with a very large driver's cab with boarded sides and an open front. This appears to have been a special concession to the Canadian climate, as it does not feature on photographs of the same tractors working in England.

Marshall tractors were built in Lincolnshire, in a factory at Gainsborough, where the company was first established in 1848. During World War I, they stopped making agricultural tractors, making a comeback in 1930 with a single-cylinder diesel-powered model which was later known as the Field Marshall (see *Great Tractors*).

The company is still in the tractor business on the original factory site at Gainsborough, building Track Marshall crawler tractors and a range of wheeled tractors based on Leyland designs which were taken over by Marshall in 1982.

(Top) Marshall 60 hp tractor photographed at the Western Development Museum, Canada.

(Above) Engine of the Manitoba Agricultural Museum's Marshall 60-hp tractor.

DEUTZ

There are very few companies in the tractor industry now which had started making tractors before World War I. One member of this exclusive group is Deutz, a company which traces its history back to the early days of the internal combustion engine and is now part of the Klockner-Humboldt-Deutz group based in Cologne.

The first Deutz tractors were built in 1907. There were two quite different versions, both designed mainly for ploughing.

(Right) Sectional diagram of the first Deutz tractor.

(Below) The first Deutz tractor.

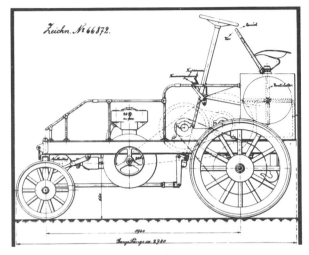

The earliest of the 1907 models was designed and built by Deutz and was exhibited on the company's stand at the 1907 DLG show at Dusseldorf. It was powered by a four-cylinder engine, which was mounted transversely and powered the rear wheels through a chain drive.

Information about this tractor suggests that it never passed the experimental stage. Problems in the field included a poor power-to-weight ratio, with the 25 hp engine proving inadequate for the 3-ton tractor. Fuel consumption during a 10 hour ploughing trial averaged about 4.25 gallons an hour and in wet conditions it was almost impossible to pull a plough.

Further development could have solved some of the problems, but the company appears to have switched its interest to a second design which was based on the Brey patents for two-way working.

The new prototype was much more complicated than its predecessor. It had four-wheel drive combined with four-wheel steering, and the upright steering column in the middle of the tractor was designed to work from seats in front and behind for two-way operation.

In the surviving photograph, the tractor is shown with four-furrow ploughs mounted at both ends. These could be lifted out of work

(Right) The first version of the Deutz tractor at work.

(Below) Deutz tractor equipped for two-way working.

mechanically, probably by means of a small winding drum with a cable.

With a 40 hp engine, the tractor was probably capable of working with four furrows in only the most favourable conditions. In order to provide more efficient pulling power, the tractor could be operated on a cable system, winding itself to and fro across the field, with the cable anchored on opposite headlands.

After this burst of tractor development activity, Deutz concentrated on other products for several years. The company became involved with tractors again in about 1917, when they designed an artillery tractor for military use. This was followed by a 40 hp tractor for forestry and agricultural work, which was built in small numbers from 1919.

GAS TRACTION BIG 4

It has been estimated that at least 700 companies have built tractors in the USA and Canada. With so many manufacturers, the industry's history is exceedingly complicated, especially as there were so many name changes, mergers, take-overs and bankruptcies during the rationalisation process.

The Gas Traction Co. is typical. The company entered the industry in about 1906 as the Transit Thresher Co., with a factory at Minneapolis. The name was changed to the Gas Traction Co. in 1908 and, in about 1910, a second factory was opened in Winnipeg to supply the Canadian market.

Several tractor models were built by Gas Traction, but they were all sold under the Big 4 name. In 1912, Gas Traction was taken over by Emerson–Brantingham, which retained the Big 4 name for a while, but discontinued the Gas Traction name.

Emerson–Brantingham was bought out by the J. I. Case Threshing Machine Co. in 1928, which was also the year that the Case company simplified its name to J. I. Case Co. Later, control of Case passed to the Kern County Land Co., and then to Tenneco, where it was joined by David

A Big 4 working in Australia with a 26 ft stripper harvester.

Brown which changed its name to Case in 1983.

The Big 4 name was appropriate. The tractors were big, and the 4 represented the number of engine cylinders. The Transit Thresher tractor had been launched with a four-cylinder engine, which was still unusual in 1906, and the manufacturers emphasised their engine design as a sales feature.

The biggest selling model weighed about 13600 lb and had 8 ft diameter driving wheels. The four cylinders were of 6 in diameter with 8 in stroke, and the output was about 35 hp. This model was built in Minneapolis and was later produced in the Canadian factory.

Canada was a big market for the Big 4 and the tractors were regularly entered in the Winnipeg trials, where they were among the top performers. A Minneapolis-built tractor won the gold medal in its class in the 1910 event and took a first and a second in the following year.

Most of the contemporary pictures of Big 4 tractors at work show them with a self-guide ploughing attachment. This consisted of a simple framework attached to the front axle, extending forwards to a small guide wheel. The guide wheel was set to run in the bottom of the previous furrow, where it remained to make minor adjustments to the steering as it followed the ploughing line. The kit was useful in the big prairie lands where the Big 4 tractors were popular and was sold as an attachment for this and other big tractors.

Grant MacEwan's book, *Power for Prairie Plows*, quotes a 1912 magazine report of a farm at Zealandia, Saskatchewan, which was operated by six Big 4s. These did all the field work on the farm's 5120 acres of wheat and flax, including the cultivations, drilling and harvesting. Each tractor pulled six 8 ft binders to harvest the wheat crop, and the magazine claimed that the six tractors did the work of 180 horses.

The Big 4 takeover deal enabled Emerson–Brantingham to move into the heavyweight end of the tractor market and they expanded the model range in a bid for a bigger share of the market. Their smallest Big 4 tractor developed 20 hp at the drawbar and they offered a top-of-the-range tractor weighing more than 10 tons and providing a 45 hp drawbar pull.

Meanwhile the demand for big prairie-busting tractors had passed its peak and sales diminished as more efficient smaller tractors gained popularity. The Big 4 finally disappeared from the market in the early 1920s.

RUMELY 'KEROSENE ANNIE'

'Kerosene Annie' was the nickname given to the first production model from the new Rumely tractor factory at La Porte, Indiana.

The engine was designed to operate on low-grade fuels, including paraffin (kerosene), which contributed to the nickname. This followed development work by John Secor, who had joined the company in 1908 to help with engine-development work.

Secor brought with him a considerable experience of engine design. Working with W. H. Higgins, a company employee, he developed a special carburettor to deal with paraffin, using a water-injection system. This was the type of carburettor which was used on Rumely engines for about 20 years and it made its first commercial appearance on 'Kerosene Annie', which went into production in 1910.

The new tractor was a heavyweight, powered by a twin-cylinder horizontal engine. This was governed to operate at between 250 and 450 rpm. Lubrication was by splash and a pump from a 10 gallon sump in the crankcase.

The massive rectangular cooling tower over the front axle, another feature which appeared on most Rumely tractors for nearly 20 years, operated with oil which was circulated by a pump.

'Kerosene Annie' underwent some design modifications after the 1910 season and emerged

'Kerosene Annie', the first production model from Rumely.

the following year as the 25–45 Model B. The new model was powered by a twin-cylinder engine of the same type as the 'Kerosene Annie' power unit. The power output of the new model, and the 9.5 × 12 in cylinder dimensions, were probably similar to the 1910 version.

The photograph, which was issued by the Rumely factory in 1910, shows the tractor equipped with a canopy roof over the engine and driving platform. At least some of the 1910 tractors were shipped without this feature.

PIONEER 30–60

Pioneer 30–60 tractors were designed for farmers with plenty of land to work and with plenty of money to spend. At $3000 in 1912, a 30–60 was a big investment, and was one of the most expensive models on the American market.

The Pioneer company had been recently established at Winona, Minnesota, and the 30–60 arrived on the market in 1911 as their first production model. It appears to have been reasonably successful at the top end of the tractor market, where volumes were never very large, and substantial numbers of 30–60s were sold in Canada.

During the 15 years or so that Pioneer was active in the tractor industry, it made several attempts to move down market, where the sales volumes were bigger and the company could expand. Several smaller tractors of around 30 hp were developed, but these were never as well known as the big models, and the company retained its up-market image.

Although the 30–60 (Plate 3) was the best-known Pioneer, it was not the biggest. This distinction goes to the 45–90, which had 9 ft diameter driving wheels and was 21 ft long. Sales were small.

The driving wheels of the 30–60 were only 8 ft diameter and the tractor weighed almost 10 tons. Its engine was a flat-four, with opposed cylinders of 7 in bore and 8 in stroke.

In many respects, the early Pioneer models were typical of the immensely heavy, strong tractors which several companies were building for the prairie farming areas. Like most of its rivals, the 30–60 had a slow speed engine, splash lubrication, and exposed gear drive with drip oilers.

But there was also at least one distinctive

feature which was years ahead of its time. Both the 30–60 and the even bigger 45–90 provided the driver with a cab for weather protection, which was complete with glazed windows. No doubt this helped to justify the extra cost of a big Pioneer.

DAIMLER 30 HP

Daimler's main bid for a place in the tractor industry came in 1911 at the Royal Show, where the company announced two new models.

At that time Daimler was claiming to be Europe's biggest motor manufacturer, with 5000 employees building cars and trucks at its Coventry factory. The tractors were a logical addition to a product range which had already established a reputation for quality engineering.

Both tractors were designed principally for export markets, as demand in Britain was still too small to support a substantial manufacturing industry. The larger model was rated at 100 hp and the smaller tractor was described initially as 30 hp, but later as 36 hp.

Daimler's 30 hp engine was a four-cylinder unit, designed to operate on paraffin. It used the sleeve valve arrangement, for which Daimler held patent rights, and which had helped the company to earn a reputation for smoothness and quietness at the expensive end of the car market.

A leather-faced cone clutch transmitted the power to a gearbox with three forward ratios and a reverse. This gave the tractor a top speed of 7 mph, with a maximum of 2.5 mph in the bottom ratio. In spite of what appears to have been a relatively high first gear ratio, the manufacturers claimed the 30 hp would handle a 10-disk plough in bottom gear.

Daimler's reputation for advanced engineering was evident in the design of the tractor. An example of this was the lubrication system, which was operated by means of a plunger pump conveniently located in the driver's cab.

Another thoughful feature was a safety device which operated when the tractor was working stationary equipment with its rear-mounted pulley. When the drive to the pulley was engaged, an automatic locking system came into operation to hold the tractor gearbox in neutral. This meant that the gears could not be accidentally engaged while the pulley was working.

Details of Daimler's tractor production are not available, but there is little evidence that the company achieved much commercial success. In 1914, the outbreak of war gave the company new priorities, making vehicles to meet military needs, including artillery tractors.

After the war, Daimler lost their interest in the tractor market and an arrangement for Foster to market a tractor in collaboration with them failed.

Daimler 30 hp.

CASE 20–40

The tractor trials which were organised in conjunction with the Winnipeg Industrial Exhibition attracted many of the leading manufacturers. The Canadian market was expanding and the trials, held annually between 1908 and 1914, brought valuable publicity for the successful competitors.

One of the most consistently successful companies in the competitions was the J. I. Case Threshing Machine Co. They took top awards in both the tractor and steam sections.

Success with steam was hardly surprising, as Case had become firmly established as the biggest manufacturer of steam engines in the world. But the company was still a small-scale newcomer in the tractor market when they started winning in the tractor classes at Winnipeg.

The 20–40 tractor won a gold medal for Case in 1913, when the emphasis of the event had changed to efficiency. Case had entered the tractor market in 1911 with a 30–60 tractor which was built for the company by an outside supplier. The 20–40 was launched as a new model in 1912

With a slow-revving twin-cylinder engine, the 20–40 was a conventional heavyweight tractor.

Case were quick to change their policy to designing lighter, more advanced models, but the 20–40 remained in production until at least 1920, to become one of the first tractors to be tested at Nebraska.

The Nebraska results provide an interesting comparison between the performance of the 20–40, which was by then an old-fashioned design, and the more up-to-date Case 22–40. The two-cylinder engine of the old tractor developed its rated power at 475 rpm, against 850 rpm for the four-cylinder engine of the 22–40.

In the drawbar section of the tests, the older model used 6.35 gallons of fuel per hour, equal to 3.89 hp-hours per gallon. The 22–40 burned only 3.84 gallons an hour, producing 6.13 hp-hours per gallon, which was one of the best performances in the 1920 tests.

One section where the older tractor scored best was wheelslip. With a weight of 13780 lb, the wheelslip was recorded as 4.76 per cent, against 13.53 per cent for the 9940-lb 22–40 tractor.

Case 20–40 tractor in an early publicity picture from the manufacturer.

INTERNATIONAL HARVESTER MOGUL 12–25

By 1910, the International Harvester Company was easily the biggest tractor manufacturer in the world and dominated the American market with a claimed one-third share of sales. This was a remarkable achievement for a company with only 5 years' experience in the tractor industry.

The tractors were built in two factories. Production in the Milwaukee factory had started in 1908, and the Chicago plant was in production from 1910. Two tractor ranges were built, with Titan models produced in Milwaukee and Moguls in the Chicago works. At that time, the company was operating two more or less separate distribution networks in the USA, and the Titan tractors tended to be sold through the original dealers, leaving Deering–McCormick outlets to sell the Moguls.

Production of the Mogul 12–25 began in 1913 and the tractor was still available in 1919. It was typical of the mechanically simple, sturdy tractors on which the company built its reputation, but with a rectangular styling which made it one of the most distinctive tractors on the market.

International described the 12–25 (Plate 4) as a medium-weight tractor in some of its American publicity, but a British advertisement of 1915 claimed it was a light tractor. The weight quoted by IH was almost 4.5 tons.

The engine was a twin-cylinder design, with a 550 rpm rated speed. It burned paraffin with jump spark ignition, and had automatic force-feed lubrication. There was a chain drive, and the

(Right) This British advertisement describes the 12–25 as a light tractor.

(Below) A Mogul 12–25 photographed at a demonstration in Scotland.

I. H. C.
Mogul Oil Tractor
12 25 H.P.

Operates on Paraffin.

WRITE FOR FULL PARTICULARS.

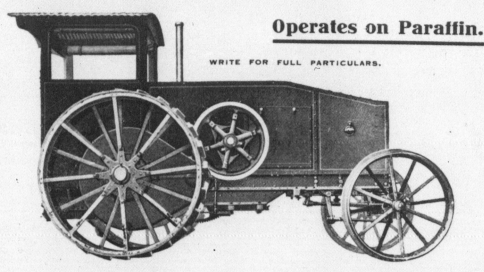

The Mogul 12/25 H.P. Oil Tractor has been designed to meet the growing demand for a light Tractor. Its total weight with all tanks filled is only four tons six cwts. It is eminently suitable for road hauling, ploughing, thrashing, harvest work, in fact any work to which a Tractor can be put. It will do the work of several men and horses, and is consequently a great money saver.

SPECIFICATION AND EQUIPMENT IS AS FOLLOWS:

POWER PLANT.	25 h.p. Two Cylinder opposed Oil Engine.
IGNITION.	High Tension Magneto. (No lamp or batteries.)
COOLING.	Water Cooled Radiator with Fan.
LUBRICATION.	Automatic force feed Oiler.
CLUTCH.	Disc type, Friction Clutch. .
TRUCK FRAME.	Heavy Steel Channels, mounted on Springs.
WHEELS.	Flat spokes, Mogul quick detachable lugs on rear and removable flanges on front.
WHEEL TRACK.	Seven feet.
SPEEDS.	Two speeds, one reverse. Top speed 4 m.p.h.; bottom speed 2 m.p.h.
PULLEY.	Fitted with standard size Pulley (friction clutch for thrashing, &c.).
MEASUREMENT.	Length, 13 ft. 3 in. ; Width, 7 ft. 6 in. ; Height, 8 ft. 4 in. Front Tyres, 6 in. ; Rear Tyres, 12 in.

INTERNATIONAL HARVESTER COMPANY OF GREAT BRITAIN, LTD.,
80, FINSBURY PAVEMENT, LONDON, E.C.

Mogul 12–25 tractor.

gearbox offered two forward ratios, with 2 and 4 mph travel speed.

Some of the more advanced features of the 12–25 tractor included the enclosed chain and sprockets of the final drive – at a time when most manufacturers left the drive exposed to dust and mud – and automotive type steering. The 12–25 was the first IH tractor to have a radiator and powered fan instead of a hopper-style cooling system.

AVERY 40–80 AND 45–65

The Nebraska test series was introduced to provide farmers with an independent assessment of some of the tractors available on the American market. During the industry's boom years, some of the new makes and models had arrived on the farm with substantial design faults, lacking an adequate parts back-up and failing to live up to manufacturers' claims for power output.

Data from the tests was – and still is – a valuable guide to efficiency. Some of the tests during the first few years of the scheme brought a quick response from manufacturers anxious to deal with faults identified by the Nebraska team.

The Avery Company of Peoria, Illinois had eight tractors tested in 1920, the first full year of the scheme. This was more than any other manufacturer and most of the results obtained were satisfactory. An exception was the 40–80 tractor, which must have caused some embarrassment when the test report was published.

Avery had brought the 40–80 into their product line in 1913, as part of their policy of concentrating on the market for big tractors. The engine was a horizontally-opposed design, with

Avery 45–65.

four cylinders which were effectively two twin-cylinder engines operating side by side.

There were two forward gears, selected by the curious sliding-frame mechanism which Avery used on most of their early tractors. The final drive was by bull gear and pinion, which were fully exposed. The cooling system was of the old-fashioned type, using the engine exhaust to induce a draught through the cooling tower, through which engine water circulated by ther-mosyphonic action.

In the Nebraska test data, the weight of the 40–80 was recorded as 22000 lb.

Some of the performance figures from the test are interesting. Water consumption was exceptionally high and, in the maximum-load belt test, the 40–80 used 18.75 gallons an hour. Only the Avery 25–50 exceeded this figure by using 25 gallons an hour.

Avery 40–80.

The fuel consumption was also high. In the drawbar test, the Avery used 11.85 gallons of paraffin an hour, which was very much higher than any other tractor tested that year. When the consumption was calculated on a hp-hour basis, the efficiency figure was poor, but by no means the worst of the year.

Perhaps the biggest surprises were in the figures for the amount of power developed. The 80 hp rating on the belt proved to be much too generous and the power actually achieved in the rated load test was only 65.73 hp. But, in the drawbar test, the Avery easily beat its 40 hp rating by achieving 46.93 hp.

After the test results were published, the manufacturers took prompt action to deal with some of the problems. The 40–80 was given a new rating and was called the 45–65 to reflect the performance achieved at Nebraska. The water consumption problem was dealt with by replacing the old cooling system with a more modern radiator and belt-driven fan.

The 45–65 remained as part of the Avery product line into the 1930s as one of the last of the old-style heavyweights.

MUNKTELLS

Munktells, now part of the Volvo organisation, began building steam engines in 1853 and later became the biggest manufacturer of traction engines in Sweden. Internal combustion engines followed in 1893 and this led to the company's interest in developing a tractor.

The first tractor built by the company, and probably also the first in Sweden, was completed in 1913. The design was influenced by the development of heavyweight models in the USA, and the Munktells tractor weighed more than 8 tons, with driving wheels almost 7 ft high (Plate 5).

An early photograph of the 1913 Munktells tractor.

One of the original Munktells tractors photographed recently near the factory in which it was built.

The most impressive feature of the tractor was the engine, which had two cylinders with a total capacity of 14.4 litres. The engine was a two-stroke design, operating on crude oil, and with a cooling system which held more than 83 gallons of water. The normal power output was 30 hp, but a maximum output of 40 hp was also quoted.

Three forward gear ratios provided a maximum travel speed of only 2.75 mph in third, with 1.7 mph in bottom gear. The final drive to the rear wheels was through exposed gearing.

Tractor No. 1 is still in existence in working order at the Volvo headquarters at Eskilstuna. It was followed by a further thirty in the same series, all built between 1913 and 1915.

Most of the tractors were bought for use on large estates in Sweden, but six were exported to six different countries, including the USSR and Argentina.

Munktells returned to the tractor market later, making them under their own name until the merger in 1950, when the Volvo BM name was used. In 1983, Volvo started making agricultural tractors under a joint development programme with the Valmet company in Finland.

BULLOCK CREEPING GRIP 35–50

The Bullock company of Chicago specialised in crawler tractors and the 35–50 was one of the most powerful models the company had produced.

The tractor had been designed and built by the Western Implement and Tractor Co., which had collapsed under financial difficulties in 1913. The 35–50 was one of the assets of the company which was taken over when the Bullock company was formed. It appears to have remained in the Bullock product range until about 1915.

With an overall length of 18 ft, the 35–50 was a big tractor and the weight was almost 8 tons. But the ground pressure under the tracks was claimed to be less than 7 lb per square inch, with tracks which were 20 in wide and had a 5 ft contact length on each side.

A feature of the track design was the pivoting action. This helped to maintain more effective ground contact over undulating surfaces.

The engine developed its rated power at 600 rpm and had four individually water-jacketed cylinders with 6.5 in bore and 8 in stroke. The manufacturers claimed the engine was easy to start because of a special trip on the magneto. This increased the armature rotation speed to give a more vigorous spark and there was a compression release valve to make the engine easier to turn by hand.

The rear section of the tractor frame carried a collection of three cylindrical tanks. The smallest of these was for petrol to start the engine and held 15 gallons. The middle-sized tank held 56 gallons of paraffin and the biggest tank contained 150 gallons of water for the cooling system.

Three forward ratios provided a maximum travel speed of 3.4 mph, with 1.77 mph in reverse. The tractor was claimed to handle an 8-furrow plough, with fuel consumption ranging between 2.5 and 3 gallons an acre ploughed, depending on ground conditions.

(Below and right) Bullock Creeping Grip tractor.

PETRO–HAUL

The Petro–Haul Manufacturing Co. of Chicago announced their new 24 hp tractor in 1914. It was based on a previous three-wheeler which perched the driver high on top of a curious rectangular framework.

Although the 1914 tractor provided a more conventional location for the driver, it still offered plenty of unusual features.

One of these was the mechanical lift mechanism for rear-mounted implements. This was operated by means of a foot pedal actuating a cam on the rear axle. As the axle turned, the cam raised a lifting arm at the rear of the tractor on which the implements were suspended. Presum-

The Petro-Haul with its mounted plough.

ably there was some locking mechanism to allow the driver to choose where and when he dropped the implement back into work.

Also unusual was the arrangement of retractable lugs on the 20 in wide rear driving wheel. There were 12 lugs which could be protruded up to 3.25 in beyond the wheel rim, or be completely withdrawn. The position of the spuds was controlled by an eccentric mechanism. This could be adjusted to protrude and withdraw the spuds at any point around the wheel. On a hard road, the lugs were withdrawn as they approached the bottom of the wheel, with the arrangement reversed when working on soil.

The Petro–Haul designers provided several features for the driver's comfort. The tractor was fully sprung, with twin leaf springs over the front axle and coil springs at the rear. There were steps and grab handles on both sides to help the driver climb aboard and a large container behind the seat to hold his tools and personal belongings.

Waukesha supplied the four-cylinder engine, which operated on petrol or paraffin. The drive to the back wheel was by means of a completely exposed roller chain and sprockets, ideally located to collect dust and mud from the wheels.

Ambitious claims were made for the tractor's performance. The ploughing rate was said to be 9 to 10 acres a day and the fuel consumption on light work, such as harrowing, was only 2 gallons for up to 25 acres. The effective pulling point was from the rear wheel axle, and the 5400 lb tractor was claimed to match the performance of many 10000 lb models.

In spite of some interesting design features, and the unusual consideration for the driver, the Petro-Haul seems to have made little impact on the market. Exactly when it vanished from the market is not clear, but it was probably not long after it was launched.

BULL

By about 1914, many more manufacturers in the USA were recognising the need for smaller, lighter tractors which more farmers could afford to buy and operate. The Bull Tractor Company of Minneapolis was one of these. Their first tractor was completed in 1913 as a pre-production prototype, and was on sale early in 1914. It was an unconventional three-wheeler, powered by an opposed twin-cylinder engine rated at 12 hp, with 5 hp available at the drawbar. It was known as the Little Bull.

Sales were encouraging at first, which probably reflects the lack of effective competition at that time and the lack of experience among farmers. The Little Bull was designed as a ploughing tractor, but its drawbar pull was quite inadequate for anything less than the easiest of soil conditions and there were some design faults which quickly earned the tractor a reputation for poor reliability.

Little Bull tractors remained on the market for about 2 years and were joined in 1915 by a new,

more powerful model which was called, not surprisingly, the Big Bull. The weight of the new model was 4500 lb, compared with 3280 lb for the Little Bull, and the engine – still an opposed twin – developed 25 hp, with 10 hp at the drawbar.

The general design of the new model was based on the Little Bull, including the tricycle arrangement with each of the wheels having a different design and function.

The front wheel steered the tractor. It was positioned on the right-hand side of the tractor so that it ran in the furrow bottom while ploughing, achieving what the manufacturers described as a completely self-steering effect.

On early versions of the Big Bull, the drive went to the big rear wheel only. This proved unsatisfactory in difficult soil conditions and, at some stage, the design was changed to put the

(*Above right*) *A Bull tractor at work.*

(*Right*) *Front view of the Bull tractor.*

power through both rear wheels to achieve more effective adhesion for drawbar work. The driving wheel followed in line behind the steering wheel in the furrow bottom.

The smaller rear wheel ran on unworked land. Most people still accept the fact that a wheeled tractor has to be tilted when ploughing because the wheels in the furrow bottom are below the level of the wheels on the unploughed soil. The Bull tractor had a cranked axle to carry the rear land wheel, with a handle to adjust the position of the axle and wheel so that the tractor could be levelled up while ploughing.

Features of the design included a force-feed lubrication system and extensive use of roller bearings. The engine was designed for easy access for maintenance and repair, with the cylinders bolted to the crankcase. The transmission provided a single ratio forwards and in reverse, with a maximum travel speed of 2.6 mph in both directions.

The Big Bull was a considerable success and, for about 2 years, it was among the top-selling tractors in the USA. It attracted the attention of the Massey–Harris company, which at the time was looking for a way to move into the tractor market. An agreement was signed under which the Bull company would supply tractors for Massey–Harris to distribute throughout Canada.

This could have been an excellent arrangement for both companies. It would have given Massey–Harris a well-established tractor to add to the company's highly successful machinery product line and it would have provided Bull with a really strong distribution outlet in the important Canadian market.

Very few tractors were delivered before the agreement came to an end. One of the problems was that the Bull company failed to deliver the number of tractors which Massey-Harris required. The Bull company was in the difficult position of relying on an outside supplier to build the tractors and the supplier decided to build tractors for another company.

Another problem was that the Fordson arrived on the market in 1917, with the Canadian Government as one of the first bulk purchasers. The Fordson quickly made most other tractors seem expensive and out of date, so that tractors like the Bull were harder to sell.

Massey–Harris went elsewhere for their tractors, while a new supplier was found to build Big Bull tractors for the Bull company. But the market for tractors like the Big Bull was past its peak and the Bull Tractor Company went out of business within about 3 years.

SANDUSKY 15–35

Sandusky tractors were named after the town in Ohio where they were manufactured by the Dauch Company. Their most successful model was the 15–35, which had a distinctive rectangular trunking from the engine compartment to the radiator.

The power unit was a four-cylinder engine designed and built by Dauch, which developed its rated power at 750 rpm. Three forward gears produced road speeds of 2, 4 and 6 mph and to stop the tractor three foot brakes were provided – one for each driving wheel, plus what was described as a 'general brake'.

The Dauch Co. made a determined effort to establish the Sandusky tractor in the British market. At least one tractor was shipped to England in 1915 and it was demonstrated at several public events including the Royal Highland and Agricultural Society trials at Stirling in Scotland.

In the ploughing section, the Sandusky was hitched up to a four furrow plough, which the judges claimed was totally unsuitable for the field conditions. In spite of this mistake, the official report had high praise for the tractor

'It is a strongly constructed and well finished machine and is very powerful. The tractor appeared to

A 15–35 Sandusky tractor.

have ample power and was well handled. With a suitable plough the observers see no reason why it should not perform good work on a large holding.

'For the thrashing this tractor took up its position very smartly, and performed the work with complete satisfaction. On the road it hauled a gross load of 6 tons with ease, its rate of travelling being considerably in excess of the other oil tractors.'

The tractor was also entered in a large-scale trial in Lincolnshire in the same year, when the engine was described as developing 35 hp with a 33 per cent reserve. The petrol consumption was estimated at 2 gallons per acre when ploughing.

Although the Sandusky remained on the American market until about 1920, it failed to attract customers in Britain and the manufacturers therefore abandoned their export sales drive.

HAPPY FARMER 12–24

In the mushrooming development of the American tractor industry during the peak period of wartime demand, companies appeared, disappeared and amalgamated with confusing frequency. The situation could be exceedingly complicated, and in some cases it is difficult to unravel the commerical relationships which developed.

One example is the link between the Happy Farmer Tractor Company of Minneapolis and the La Crosse organisation of La Crosse, Wisconsin. Both sold apparently similar versions of the same 12–24 tractor design and, for a while, the tractor was sold under both the Happy Farmer and the La Crosse brand names in the same advertisements.

The Happy Farmer company was the first to market the 12–24 model (Plates 6 & 7) in 1915. It was a three-wheel design, with a single wheel at the front and designed with most of the weight over the driving wheels at the back. The engine was a water-cooled, horizontal twin-cylinder, and the tractor was built around a steel tube instead of a steel girder frame.

According to C. H. Wendel's *Encyclopedia of American Farm Tractors*, the La Crosse Implement Co. also started marketing the 12–24 in 1915 and, in 1916 a new company, the La Crosse Tractor Co., was formed and this apparently took over the former interests of both the Happy Farmer and the La Crosse Implement concerns.

Happy Farmer and La Crosse 12–24 tractors, together with a smaller 8–16 model, sold in large numbers. The design was simple, with a single forward gear ratio, and the tubular chassis presumably helped to keep the production cost competitive.

In 1919, an improved version of the 12–24 was introduced. This was available with a single front wheel, called the Model F, or with a full-width front axle when it was called the Model G.

A La Crosse G was submitted for test at Nebraska in 1920, where it recorded almost 25 hp on the belt and 13.6 hp in the drawbar test, with a maximum pull of 2155 lb.

In 1921, the La Crosse company was taken over and tractor production came to an end.

LINE-DRIVE

One of the problems for salesmen in the early days of the tractor industry was that hardly anyone knew how to drive a tractor, but almost everyone on a farm could control a horse. It seemed logical, therefore, at least to some designers, to build a tractor which could be handled in very much the same way as a farm horse. Several companies decided to give the idea a chance, including Fowler in England and the Line-Drive Tractor Company of Milwaukee in the USA.

The Line-Drive version arrived on the market in about 1915 and disappeared again about 3 years later.

All the normal controls, including steering, clutch operation and braking, were operated through the lines, which the driver held in his hands as if he were working with a team of horses. The lines could be long enough to reach the seat

of a trailed implement, or the driver could use the seat over the pair of rear wheels.

The Line-Drive sales effort was directed firmly towards the farmer who was accustomed to using horses and who might be reluctant to make the change to something as unfamiliar as a tractor.

'Anyone that can handle a team of horses can drive this tractor', the company claimed. 'From a seat on a binder, on the top of a load of hay, on the seat of a manure-spreader or gang plow, or from any place where the lines may extend.'

Exactly how the lines controlled the tractor mechanism is not clear, but some of the mechanical details are still available. The engine was rated at 25 hp and this gave 15 hp at the drawbar,

The Line-Drive tractor.

which the manufacturers claimed would handle a two-or three-furrow plough.

The tractor weighed almost 2.5 tons, with nearly all of the weight over the 66 in diameter driving wheels.

One of the complications resulting from the Line-Drive's unconventional layout was the steering system. The engine, transmission and wheels were all assembled as one unit, which was mounted on a turntable attached to the steel tube which acted as the tractor's main frame.

To change direction, the driver presumably tugged on the appropriate line and the engine and wheels swivelled on the turntable to face towards left or right.

In spite of a determined sales effort, both in the USA and in some potential export markets, demand for the Line-Drive was disappointing. Even the argument that the driver was isolated from the vibration and much of the noise of the engine failed to win sufficient customers.

The Line-Drive was still being promoted in 1917 but disappeared from the market soon afterwards.

TOM THUMB 8–18

For $600 in 1915, the Tom Thumb Tractor Co. of Minneapolis provided plenty of mechanical ingenuity, but not much pulling power.

The engine was a horizontally-opposed twin cylinder, with 5.5 in bore and 6 in stroke. This, the makers claimed, developed 18 hp at 625 rpm. But the drawbar power was only rated at 8 hp,

suggesting an above-average power loss and indicating that the single-track propulsion system may not have been such a good idea.

Using one track meant considerable mechanical complications, not least in the steering system. When a change of direction was required, the driver lifted a hand lever to raise the front

On the Tom Thumb tractor, the fuel tank was beneath the driver's seat.

section of the track. The lifting mechanism was powered from the engine, but the detail of how this was achieved is not clear. With just 12 in of ground contact at the rear end of the track, there was sufficient grip to continue driving the tractor forward while it was steered.

Another curious feature of the design was the attachment point for trailed implements, which pulled from the centre of the rear track sprocket. The makers claimed this was an advantage, as it avoided putting the pulling stresses through the

tractor frame. Exactly why it was preferable to put the stresses through the main drive sprocket is not obvious.

The benefits of the single-track idea, the manufacturers claimed, were the narrow working width to suit rowcrop working, and the low ground pressure to avoid soil compaction. The track had a ground contact area of 6 square feet to carry the tractor's 3000 lb weight.

Tom Thumb also built a 12–20 version, powered by a four-cylinder engine, and using the same single-track drive system. There is little evidence that the idea was a commercial success and the Tom Thumb soon faded from the market.

Rumely's All-Purpose three-wheeler.

RUMELY ALL-PURPOSE

The All-Purpose motor plough was Rumely's first serious attempt to break into the lightweight end of the market and to meet the growing demand for less expensive tractor power.

Two models were announced in 1915, both based on a highly unconventional three-wheel design. The smaller model was designed to work with a mid-mounted two-furrow plough and was powered by a 16 hp, four-cylinder engine operating on petrol or paraffin.

A 24 hp engine was used on the three-furrow model, which had a lengthened frame to give extra space in the wheelbase to accommodate the additional furrow.

The All-Purpose was designed to operate as a two-way machine. For ploughing, the tractor was driven with the two large wheels at the front and the single steering wheel was at the rear. With the plough unit removed, the tractor worked in the reverse direction, with the steering wheel at the front and the driving seat swung round to the other side of the vertical steering column.

As a motor plough, the design allowed all three wheels to run on unploughed land. There was a single, extra-wide driving wheel and the other front wheel was an idler.

The smaller version of the All-Purpose was not especially cheap at $750 in 1916, but it was built with typical Rumely attention to quality, including the use of roller bearings throughout.

Another feature which helped raise the manufacturing cost was the lift mechanism for the plough unit. This was controlled by a foot pedal near the driving seat and was powered from the engine through a chain drive.

The All-Purpose models arrived as the motor-plough market was developing in the USA and they survived until about 1919.

EAGLE 16–30

The 16–30 (Plate 7) was one of the most successful tractors produced by the Eagle company. It arrived on the market in 1916, when demand for tractor power was booming, and it was still available 15 years or more later in the early 1930s.

It was the 16–30, plus the smaller 8–16 model introduced in the same year, which established Eagle as one of the more substantial tractor manufacturers in the USA. Both tractors arrived at a time when sturdy, but basically simple, tractor designs were finding a growing demand and the new Eagle models suited this requirement.

At this stage, the company was still making its own engines, although a policy change in the 1930s brought in Hercules and Waukesha power units. The 16–30 was powered by a two-cylinder, horizontal engine developing its rated power at 500 rpm. The engine was unusual in having equal bore and stroke dimensions of 8 in. The transmission included two forward gear ratios and a reverse, with exposed final drive.

During the 1920s, the Eagle factory at Appleton, Wisconsin, expanded its tractor range with new models of various sizes (eg the 12–25 shown in Plate 8), as well as developing the original 16–30 tractor. Throughout this period, the company continued with most of the 1916 design features, including the sideways-facing radiator, the slow speed, horizontal engine and the steel frame construction.

By the time completely new designs began to appear in the product line in the 1930s, the old models were looking extremely dated. Some of the old-style designs were still on the market in about 1936–37.

Eagle tractors were still listed in 1940, but production came to an end soon after this date when the company was sold.

Eagle tractor.

WALLIS CUB JUNIOR

The Wallis Tractor Company made history in 1913, when they introduced the Cub tractor with a one-piece, rolled steel frame to support the main working components.

This was an important development which influenced the future design of most tractors throughout the world. It formed an immensely strong, rigid structure to which other components could be attached and it helped to simplify the manufacturing operation. The frame also provided a dirt and wet-proof underside to protect the mechanism.

On the Cub tractor, the curved frame structure protected all the main working components, apart from the final drive. The traditional external pinion-and-ring gear was retained on the Cub in its usual fully-exposed form.

When the smaller Cub Junior model went into production in 1916, the designers had completed the logical development of their one-piece frame by enclosing the final-drive mechanism.

The curved shape of the frame, which was made of rolled steel plate, remained a distinctive Wallis feature and survived the Massey–Harris takeover in 1928, when it continued in the Pacemaker and Challenger models introduced in 1936 (see p.116).

Power for the Cub Junior was produced by a four-cylinder engine rated at 25 hp and delivering 13 hp at the drawbar. The single front steering wheel, topped by the company's bear emblem to show the driver which way the wheel was turned, provided a tight turning circle which helped to impress observers at the 1917 tractor trials in Scotland.

'There is no denying that it was skilfully handled and easily turned at the headland in masterly style' [said the report of the trials in *Implement and Machinery Review*].

'The Wallis Cub Junior was by far and away the speediest machine on the ground, and, trailing a self-lift J. I. Case 2 furrow plough with disc coulters cutting furrows 14 in wide, it finished its plot in 44 minutes.

Later versions of the Cub Junior included more engine power and a four-wheel model.

Wallis Cub Junior showing one-piece curved steel frame.

CLAYTON

British farmers showed little interest in the idea of using crawler tractors before the start of World War I, and, up to this time, there was only a small amount of tracklayer development for agricultural purposes in Britain.

The situation changed during the war and this was probably very substantially influenced by the publicity given to the successful performance of tanks operating with the British army. For a few years before and after the end of the war there was a sharp increase in the number of British companies building crawler tractors, and more imported makes were also available.

Clayton and Shuttleworth, a famous name in the steam-engine business, built their first crawler tractors (Plate 10) in 1916 and at least some of their production during the next 2 years was against Government-backed orders to supply tractors for the ploughing-up campaign.

At first, the tractors were described as 35 hp, but this later became 40 hp. The tractors were powered by a Dorman four-cylinder engine of 6.3 litres, governed to 1000 rpm maximum.

'. . . a woman can handle it, with plough attached', the manufacturers rather ambiguously claimed, referring principally to the steering system. This was operated by a steering wheel, with large cone clutches to engage the drive to the main track sprockets on each side. For tight turns, there was also a foot-operated brake to lock either of the tracks separately.

Two forward gears provided travel speeds of 1.75 and 4 mph, with up to 3 mph available in reverse. The maximum drawbar pull in bottom gear was claimed to be 2 tons, with a tractor weight of 2.8 tons. The 1918 price was £650.

Clayton and Shuttleworth kept the tractor in production until the mid-1920s, when it disappeared from the market temporarily. It was reintroduced with some minor alterations in 1928, but this was also a short-lived arrangement. Clayton and Shuttleworth was taken over by Marshall of Gainsborough in 1930 and further tractor development was concentrated on the single cylinder Marshall diesel.

Clayton tracklayer.

INTERNATIONAL HARVESTER MOGUL 8–16

In 1914, International Harvester introduced two highly successful new models. The Titan 10–20 (see *Great Tractors*) was built in the Milwaukee factory and the Mogul 8–16 (Plate 11) came from Chicago.

The new Mogul was the smallest model in the International range and also the least sophisticated. Everything about the tractor was designed for simplicity and it built up an excellent reputation for sturdy reliability.

An example of the mechanical simplicity was the engine. This was a single-cylinder unit which developed its rated power at only 400 rpm, and used a hopper cooling system.

The transmission was similarly basic, with as little as possible to give trouble. There were two forward ratios, providing a leisurely top speed of 2 mph, with a single chain final drive.

A single, straight rod, almost as long as the tractor, linked the steering wheel at the one end with a worm gear at the other end. A pulley was provided for belt work and this was incorporated into the flywheel.

The 8–16 remained in production until 1917. It was joined in 1916 by the last new tractor to carry the Mogul name, the 10–20. The new model provided the extra power which many farmers required, but was otherwise simply a slightly bigger version of the 8–16, with the addition of mudguards over the rear wheels as a distinguishing feature.

(Inset) The International Harvester Mogul 8–16.

An International Harvester Mogul 8–16 in a photograph taken during the ploughing campaign in wartime Britain.

EMERSON–BRANTINGHAM 12–20

The Emerson–Brantingham company of Rockford, Illinois, bought its way into the tractor market by taking over several established manufacturers in 1912. The purchases included such well-known companies as Reeves and the Gas Traction Co. with its successful Big 4 models.

With the help of these new acquisitions, Emerson–Brantingham was firmly established in the market for big tractors, but needed additional models to compete for a share of the growing demand for smaller models.

The first new model designed by the company went into production in 1916. It was a three-wheel tractor with a power rating of 12–20 hp. Two further models with the same power rating went into production in the following 2 years.

The second and third 12–20s were both four-wheel tractors. One of these was announced in 1917 and is the version shown in the photograph (see also Plate 12). It weighed more than 2.5 tons and the design included several features which were already becoming dated.

Fully-enclosed final drives were appearing on many new tractors by 1917, because of the high wear rate and lubrication problems of the exposed mechanisms of the type used by Emerson–Brantingham. The 12–20 engine (Plate 13), with four cylinders in two separate blocks, was also beginning to look old-fashioned.

The third 12–20 arrived in 1918, using the

Emerson–Brantingham 12–20

Emerson–Brantingham final drive.

same engine design, but with a fully-enclosed drive mechanism. This tractor was also lighter than its immediate predecessor, but developed slightly more power through an increase in the engine speed. In the Nebraska tests, the last of the three 12–20s developed a creditable 24.9 hp in the maximum load test.

Both the 1917 and the 1918 versions remained in production until the company was bought out by J. I. Case in 1928.

A batch of twelve of the 1917 model 12–20s was shipped to Britain, arriving in 1918. The tractor illustrated is believed to be the sole survivor from this shipment and is now preserved, in beautifully restored condition, at the Hunday National Tractor Museum in Northumberland.

BATES STEEL MULE MODEL C

There was a series of Steel Mule tractors of various designs, but it is the Model C which most people recognise because of its curious appearance.

The tractor was first produced in about 1916 at Joliet, Illinois, and appears to have remained in production until 1919. In spite of its unconventional design, it enjoyed a brief popularity and several of the tractors have been preserved.

One of the features which gave the Model C its unique styling was the rocket-shaped structure at the top. This consisted of a front section, which was the radiator with a pattern of holes at the front. At the opposite end was the cylindrical fuel tank, with a middle section which was simply a shaped cover to protect the engine.

Even without the top structure, the tractor was still highly unconventional. The tractor was

driven by a single track which was centrally positioned at the rear of the machine. This did little for the stability of the Steel Mule on sloping ground and, to help keep the tractor upright, the front wheels could be set wide apart, as in the photographs.

The photographs show the driver seated close to the tractor, with an assistant sitting on the rear of the plough. The driving position could also be set back far enough for the driver to sit on the implement being pulled. This involved extending the steering column and the main hand controls by up to about 8 ft.

The power unit was an Erd four-cylinder petrol engine, rated at 30 hp which was devel-

Close-up of the Bates Steel Mule working in Britain.

A Bates Steel Mule taking part in a competitive trial in Britain.

oped at 900 rpm. There was a chain drive to the single track, and another chain and sprockets formed part of the steering mechanism, transmitting movement in the steering column to a shaft at the side of the main frame.

When the Model C tractor was demonstrated in England in 1917, its performance was impressive.

'Farmers present were loud in its praise [said one magazine report]. 'It operated in a field of lucerne which had not been ploughed for about six years, and conducted its work unfailingly, leaving in its wake three neat and even furrows.'

CLEVELAND MOTOR PLOW

The Cleveland Motor Plow Co. was formed in January 1916 at Cleveland, Ohio, by Rollin H. White and some of his associates. White, a member of the steam car family, had started development work on farm tractors in 1911. In about 1914, he and an assistant had developed the controlled differential steering system for crawler tracks.

This replaced the usual steering clutches and hand levers by a system of gears and a steering wheel. Turning the wheel applied a braking effect on the gears on one side of the tractor, leaving the gears on the opposite side free to continue driving the track. This gave a positive and precise control over the steering radius and retained the familiarity of the steering wheel.

The controlled differential system was a significant breakthrough in crawler tractor design and it encouraged White to set up a company to build tractors incorporating the idea.

By August 1916, the first tractor was being built and it was followed by more, in increasing numbers, before the end of the year. The tractor was called the Model R Cleveland Motor Plow, and it was a small tracklayer designed particularly for areas of intensive crop farming, such as California.

A Weidley engine was used, rated at 18 hp on the belt and producing 10 hp at the drawbar. It was a four-cylinder unit developing its rated power at 1200 rpm. Only one gear ratio was provided for both forward and reverse. The price was $1185.

At the front of the tractor was a large transverse leaf spring. There was also a front-mounted

An early Cleveland tractor with its 'geared to the ground' publicity message.

pulley. This was placed transversely so that the driving belt was at a right angle to the tractor. It was an arrangement which was retained on several of the early models produced by the company and must have created problems when lining up to achieve a suitable belt tension.

In the following year, the first model was redesigned to improve the efficiency and reliability. The changes included improved bearings in the lower track wheels and fitting a new type of air cleaner on the engine. The improved model was called the H. It offered an extra 2 hp at the drawbar and the price was raised to $1385.

Another important development in 1917 was a changed name for the company. The new title was 'The Cleveland Tractor Co.' and, for the remainder of the year, the tractors were simply called Cleveland. 1918 brought another name change, with Cletrac being introduced as the trade name for the tractors. This was the name which survived until after the company was taken over by Oliver in 1944.

CASE CROSSMOUNT SERIES

It was the Crossmount series which helped to bring Case into prominence as one of the leading companies in the American tractor industry.

The J. I. Case Threshing Machine Co. had already become firmly established as the USA's biggest manufacturer of agricultural steam engines and threshing machines. In this situation, the decision to diversify into the tractor market in 1911 showed considerable foresight.

For a few years after 1911, the tractor market in North America was dominated by the demand for big, heavyweight models and Case concentrated on this type of tractor. But Case again showed good commercial judgement when it began building smaller models to match the changing needs of the market. This happened in 1913 when a medium-size 12–25 tractor was introduced.

The first of the Crossmount models arrived in 1916. The term 'Crossmount' was adopted because of the transverse engine arrangement, which spread through most of the Case range as new models were developed until about 1929.

Although Case was not the only manufacturer to use a transverse engine design, it was the Case range which made a special success with this arrangement, in a series of tractors which were particularly distinctive in appearance (Plate 15).

There are several theoretical advantages in using a transverse engine. The power line from the end of the crankshaft to the back axle can be simple and direct, and this also applies when a side-mounted belt pulley is used. There might also be an advantage in avoiding lubrication problems in the crankcase when the tractor is working up and down steep slopes.

To what extent these factors may have influenced the success of the Case tractors is not clear. Probably of much greater significance is the fact that the tractors were generally well designed, with features which helped to put Case several years ahead of many competitors.

Case used four-cylinder engines throughout the range while most of their competitors were still providing two cylinders. Transmission components were enclosed for protection from dust and dirt when exposed bull gears were still standard practice for some tractor companies.

With their advanced features, the Crossmount tractors were also sturdy and heavy enough to earn a reputation for toughness and good pulling performance, and several models in the range easily exceeded their official power ratings when tested at Nebraska.

The biggest of the Crossmount models was the 40–72, which weighed more than 9 tons. At Nebraska, it achieved almost 50 hp at the drawbar and more than 90 hp on the belt. The 40–72

Transverse engine layout of a Case Crossmount tractor.

The "Last Word" in Farm Tractors

A British advertisement for the Case 22–40.

was launched in 1923, but remained in the range for only 2 years.

More typical of the Crossmount tractors were the smaller models, such as the 15–27, available between 1918 and 1925, which developed a maximum of more than 31 hp in several of the belt tests at Nebraska.

During the period when the Crossmount tractors were on the market, the Case share of the American market increased sharply from being very much a minority make to one of the top four tractor manufacturers in the USA. This was achieved at a time when scores of manufacturers disappeared into bankruptcy or mergers.

GARRETT SUFFOLK PUNCH

The success of the internal combustion engine on the farm brought increasing problems for companies building agricultural steam engines.

By the beginning of the twentieth century, the production of steam traction, ploughing and portable engines had become big business, with British and American manufacturers dominating the market. The steam engine companies responded in various ways to the growing threat of the tractor, providing an interesting example of the impact of competition from new technology on an established industry.

Some of the steam engine companies tried to ignore the arrival of the tractor and simply went on building the traditional types of agricultural steam engine for a dwindling market. Many of these companies eventually went out of business or were taken over.

Others in the industry realised, sooner or later, that there was a bleak future for steam power in agriculture and decided to join the opposition as tractor manufacturers. There were several companies which made the switch to tractors successfully, including the J. I. Case Threshing Machine Co., once the world's largest manu-

Garrett Suffolk Punch steam tractor working with a four-furrow plough.

facturer of agricultural steam engines and later one of the most successful tractor companies.

A third response to the tractor was an attempt to redesign the steam engine in order to make it more competitive and to attract new customers. Several companies in Britain and the USA tried this approach and, although none of them achieved a commercial success, some of the steam 'tractors' they produced were extremely interesting.

An example is the Suffolk Punch, built by Garretts in their factory at Leiston, Suffolk, in 1917 and 1918. Garretts called the Suffolk Punch an 'Agrimotor', a name they had previously used for a pedestrian-controlled motor plough which they had built in small numbers in 1913 and 1914.

The Suffolk Punch development project started in 1915 and was linked to the design of a new steam waggon for road haulage. The waggon was built to take a 5 ton payload and was equipped with a conventional boiler placed so that the firebox was at the forward end. There was a superheater to help improve operating efficiency.

The agricultural version shared the same basic design with the driver placed right at the front of the vehicle. Both versions had fully-sprung axles, with Ackermann steering from an upright steering column and wheel. The rear wheels of the Suffolk Punch were 5 ft in diameter and this gave extra ground clearance, which allowed space for an underslung water tank holding 185 gallons to be tucked around the axle. The final drive was by roller chain and sprocket.

Garretts had designed the Suffolk Punch for ploughing by direct traction, which was unusual in Britain with steam power. The first tractor taken went to Foulness Island to work with a four-furrow Ransomes plough on old pasture with heavy soil. The conditions were difficult, but the steam tractor performed satisfactorily, using 384 lb of poor quality coal and evaporating 2200 lb. of water for the 2-acre trial.

In these conditions, the cost per acre was

Garrett Suffolk Punch steam tractor showing the rear-mounted boiler.

estimated at 6s.3½d (31.5p) including wages for the driver and the ploughman at £1.15s (£1.75p) each for a week's wages. The working rate was 0.675 acres an hour. On easier land, the cost per acre was 4s.10d (24p).

Compared with horse-ploughing, or working with a double-engine cable set, the costings for the Suffolk Punch appeared to be encouraging. Unfortunately, the figures were unrealistically optimistic. They were based on working 10 hours a day for 50 weeks a year, which is an improbable situation for a ploughing engine, and there was no allowance for the cost of supplying the tractor with water.

Eight Suffolk Punch tractors were built, with the last one finding a customer in 1920. The unconventional design failed to match the simplicity and low cost of tractors such as the Austin and Fordson, and consequently the project was abandoned.

SAWYER–MASSEY 11–22

The Massey family's very considerable influence in the Canadian farm machinery industry started in 1847. That was the year when Daniel Massey began making implements on a small scale at Newcastle, in what is now Ontario.

His company expanded rapidly with the development of Canadian agriculture and, in 1879, they moved to a new factory in Toronto.

In 1891, the Massey and the Harris organisations between them controlled more than 50 per cent of the Canadian farm machinery market. That was when they merged to form Massey–Harris.

During the following year, the Massey family bought a substantial interest in the Sawyer company of Hamilton, Ontario, manufacturers of agricultural steam engines. The company name was changed to Sawyer–Massey Co. and the product range was expanded.

In 1910, Sawyer–Massey added gasoline tractors to their list of products and, in 1911, they were one of the few companies with entries in both the gasoline and steam sections of the Winnipeg trials.

Sawyer–Massey concentrated on heavyweight tractors initially, but moved downmarket in about 1917 when the 11–22 tractor (Plate 14) was announced. This was their smallest model and it was probably the most popular until the company went out of tractor production in 1922.

An interesting design feature of all Sawyer–Masseys, including the 11–22, was the way the weight of the engine was used to help improve wheelgrip.

All the tractors were powered by four-cylinder engines of solid, heavy construction and the designers arranged the layout so that the engine weight was as far back as possible, almost over the rear axle.

This left plenty of space in front of the engine, with little to occupy the area between the radiator and the cylinder block. In theory, the weight distribution should be an advantage, although the low power ratings for drawbar horsepower do not suggest any special efficiency advantage.

Another result of the weight distribution could be a tendency for the front of the tractor to lift if a plough or cultivator should meet an obstruction, and a low hitch position would be needed to overcome the problem.

Plate 1 Ivel tractor.

Plate 2 Hart-Parr 'Old Reliable'.

Plate 3 Pioneer 30-60 tractor.

Plate 4 Mogul 12-25 tractor.

Plate 5 Munktells tractor.

Plate 6 A Happy Farmer tractor at the Ontario Agricultural Museum.

Plate 7 Happy Farmer tractor.

Plate 8 Eagle 12-25 tractor.

Plate 9 Eagle 16-30 tractor.

Plate 10 Clayton tractor photographed in the Hunday Museum, Northumberland.

Plate 11 International Harvester Mogul 8-16

Plate 12 Emerson-Brantingham 12-20

Plate 13 Engine detail of the Emerson-Brantingham.

Plate 14 Sawyer-Massey 11-22, built and preserved in Canada.

Plate 15 Case Crossmount 22-40 model.

ONCE-OVER 15–25

The Once-Over tiller was designed and manufactured by the Scientific Farming Machinery Co. of Minneapolis, but the marketing arrangements appear to have been handled by the Once-Over Tiller Corporation.

The tiller had been developed to do a complete sequence of cultivations in a single operation. The mouldboards of the plough lifted the soil, while the steel-toothed rotors broke it down to a seedbed tilth. The result was a big saving in time and cost, the designers claimed, with a better tilth because the soil was pulverised right through the ploughing depth.

In its first version, the Once-Over was an attachment for a single-furrow plough, which was usually horse-drawn. The rotor attached to the plough frame and was powered to turn at 500 rpm by a small air-cooled engine, which was part of the Once-Over kit.

Presumably the attachment was a success. It arrived on the market in about 1916 and, soon afterwards, the manufacturers began development work on a self-propelled version. The prototype was completed in 1917 and the production version was on the market in 1919.

The tractor unit, which could be used for other purposes when the plough was removed, was powered by a four-cylinder engine developing 25 hp. This was a Buda petrol–paraffin unit which was provided with a 19 gallon fuel tank. The layout of the tractor, with the weight of the engine and fuel tank over the driving wheels, was claimed to provide 3000 lb of drawbar pull.

Some of the design features were advanced. An example was the electric starter motor, with a cranking handle provided in case of difficulties.

The drive from the engine to the rotors came from a power-take-off at the rear of the power unit, with a long drive shaft. The drive shaft also provided the power to lift the plough out of work, operating through a series of cables to lifting points on the plough frame.

Another ingenious feature was the steering system. For normal working, the steering wheel operated through a rack and pinion to give a turning cycle with 20 ft radius. In confined spaces, the driver could release a locking device which allowed the single rear wheel to caster freely. He could then lock one of the front wheels so that the tractor turned with a pivoting action in an 8 ft circle.

The Once-Over was one of several attempts which have been made at various times to combine two tillage operations in one machine. A third operation was added in 1922, when a gas injection version of the Once-Over was announced.

This was designed to release carbon dioxide into the soil as it was being cultivated. According to the manufacturers this would release 'phenomenal' quantities of plant nutrients and make the use of chemical fertilisers unnecessary on many soils.

The theory was based on the idea that almost every soil contains large amounts of the nutrients required for plant growth, most of which are in chemical forms which the plant is unable to use. When carbon dioxide is injected into the soil, it combines with the moisture which is present to produce carbonic acid. The acid then reacts with other substances in the soil and the plant nutrients are released in the process.

Gas for the injection process was available as an industrial by-product and, when the Once-Over system was demonstrated, it was using gas collected from a blast-furnace chimney.

In order to assist the injection process, exhaust gases from the Once-Over engine were mixed with the industrial waste to provide sufficient pressure for a thorough mix. The carbon dioxide supply was carried in a container on the back of the plough, injecting through specially-modified tines on the two rotors, which were hollow and drilled to give a series of holes for the gas to escape.

What scientific evidence the Scientific Farming Machinery Co. had to support their gas injection theories is not clear and there is no evidence of a big demand for Once-Over cultivators from farmers anxious to take advantage of the gas injection system. Production of the self-propelled unit apparently came to an end in about 1923.

(Overleaf) The Once-Over tractor with combined plough and powered cultivator mechanism.

CANADIAN

Canadian farms have provided a traditional export market for most of the world's major tractor manufacturers, but there have been surprisingly few successful attempts to establish independent companies to build tractors in Canada.

The Alberta Foundry and Machine Company decided to become a tractor manufacturer in 1918 and announced their first model the following year. It appears also to have been their last model.

The factory was at Medicine Hat in Alberta, which is a convenient location for distribution throughout the prairies. During World War I, the company had made ammunition for the Canadian Government, with a demand for all they could produce. The end of the war also meant a sharp reduction in the need for shells and the Alberta Foundry and Machine Co. looked around for alternative products to keep their factory busy.

Tractors were big business at the time and the company decided to design a tractor which could compete against imports in the prairie provinces. The new model was patriotically called the Canadian.

At a time when tractor design was making significant progress, some of the features of the Canadian were beginning to look out of date. The horizontal two-cylinder engine had already lost much of its earlier popularity as more manufacturers specified four-cylinder in-line units.

By 1919, thousands of Fordsons were demonstrating the advantages of doing away with a separate main frame, but the designers at Medicine Hat preferred to follow tradition, by arranging the various main components along a single beam which linked the front and rear axles.

The most unusual feature of the Canadian was the extensive use of hardwood in its construction. The frame was a 9 in square length of timber and wood was also used to form the rectangular spokes of the rear wheels.

Using wood in place of metal may have helped to reduce the manufacturing cost, but apparently it caused problems for some of the customers. In the dry conditions of a prairie summer, some of the wood tended to shrink and crack and this may have helped to discourage potential buyers.

Another discouragement was price. As Henry Ford led the price war, $1200 for the 28 hp Canadian began to look excessive. According to one report only about forty Canadian tractors were sold during a period of about 2 years before the manufacturers abandoned the project.

The main frame of the Canadian 14–28 was made of hardwood.

ALLIS–CHALMERS 6–12 GENERAL PURPOSE

The success of the universal type of motor plough in the USA attracted more manufacturers into the market with versions of their own. Allis–Chalmers was one of the companies competing for sales, with the 6–12 General Purpose model which was announced in 1918 and was on the market early the following year.

Design work for the 6–12 had started in 1915, soon after Allis–Chalmers introduced its first-tractor, the 10–18 three-wheeler. Although the 10–18 was not a success, the company continued with the development programme for the motor plough.

The basic design was mechanically simple, with the emphasis on economy. At that time, there were more than 25 million horses and mules at work on American farms, used mainly on smaller acreages, where there was little money available for changing to tractor power. These farms were an obvious target for the motor ploughs, which were designed to work with many of the items of horse-drawn equipment which were already available on most farms.

Allis–Chalmers used a Le Roi four-cylinder engine for the 6–12. This developed its rated 12 hp at 1000 rpm, and it helped the motor plough to achieve fuel consumption figures which were among the lowest recorded during the first year of the Nebraska test series.

In some conditions, the 6–12 was able to handle a two-furrow plough, but there was also a demand for more power. Allis–Chalmer experimented with the idea of linking two 6–12 tractor units together in order to double the pulling power. The idea was presumably impractical as it did not go into production.

The biggest barrier to sales was price. At $850, it was too expensive for most of the farms it was designed to suit, while farmers who could afford the price were often looking for more power and performance. The Fordson was slightly cheaper

An Allis–Chalmers General Purpose tractor with a cultivator attachment.

(Overleaf) Ploughing with the Allis–Chalmers General Purpose tractor.

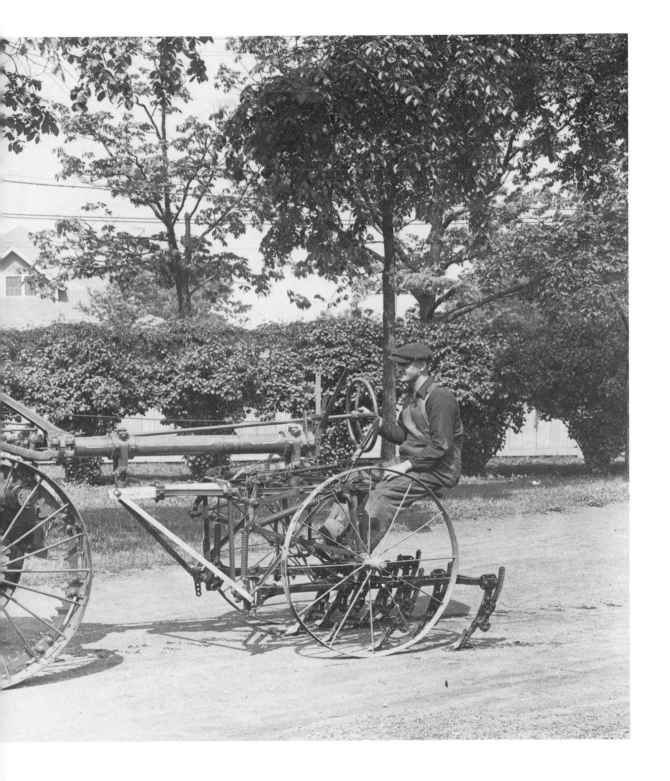

than the Allis–Chalmers and the gap widened as Ford reduced his prices to increase his sales.

Allis–Chalmers stopped building the 6–12 in 1922, after about 700 had been built. The last 200 were sold off at drastically reduced prices, with some going for as little as $295 in 1923. Allis–Chalmers moved out of the market just in time, as the more practical and versatile rowcrop tractors finished off the motor plough market within 2 or 3 years.

RUMELY OIL PULL 12–20

The design of Rumely Oil Pull tractors was noted more for being traditional than for innovation. The 12–20 Model K, in production from 1919 to 1924, was a typical example.

Although the 12–20 (Plate 16) was a small tractor by Rumely standards, it was still very much an Oil Pull in its general design. It was built on a steel girder frame, with the familiar rectangular cooling tower over the front axle.

The engine was a smaller edition of the typical Rumely twin-cylinder design, developing its rated power at a leisurely 560 rpm. The gearbox made no concessions to progress, with just two forward ratios to cope with all work conditions. At almost 3 tons the Rumely was heavier than many of its more powerful rivals.

In 1920, the Model K was one of four Oil Pulls submitted to the University of Nebraska test station. All four tractors showed that there was a considerable reserve of power available beyond the nominal rating and, for the 12–20, the maximum hp on the belt was recorded as 15.87 hp.

Another feature of the Nebraska results was an exceptionally good record for fuel efficiency in the belt tests. At the rated load, the power output was 10.82 hp-hours per gallon. This was the best result of more than sixty tractor tests carried out at Nebraska in 1920. It compared with an average result of about 7, and a few tractors produced around 5 hp-hours per gallon of fuel.

The Model K reached the end of its production run in 1924, but it was certainly not the end of the old style Oil Pulls. The same type of engine, and a basically similar styling, was still on the market when the company was taken over by Allis–Chalmers in 1931. One example of the mid-1920's Oil Pull Models, which featured pressed steel chassis side members was the Model R 25–45 (Plate 17), which was built from 1924 to 1927.

A British-built Austin tractor.

AUSTIN

British manufacturers have had considerable influence and success in France. One early example of this is the Saunderson Universal, which was manufactured in France as the Scemia through a licence agreement.

Another French success was the Austin, which achieved excellent publicity for its performance at the international trials at St Germaine en Laye in 1919. Herbert Austin, later Sir Herbert, bought a factory in northern France, where the tractor could be manufactured for the French market, and this proved to be a successful venture, operating from 1920 until the early 1930s and, at times, exporting completed tractors back to Britain.

The Austin tractor (Plate 18) was announced in Britain early in 1919. It was built at the big Austin car factory in Birmingham, using a four-cylinder engine originally developed for a new car. The engine was available in petrol or

petrol–paraffin versions, developing a maximum of 27 hp at 1500 rpm.

At the Society of Motor Manufacturers and Traders trials in Lincolnshire, the dynamometer tests on the petrol-engined Austin achieved a maximum figure of 26.5 hp on the belt, compared with 23.7 hp for the paraffin version. The maximum drawbar pull recorded in the tests was 1820 lb.

Although the Austin performed well in various tests and trials, it faced the inevitable problem of competition from imported Fordsons. The 1921 price of the Austin was £360, or three times as much as the slightly less powerful Fordson.

The manufacturers maintained an apparently optimistic view of the situation. A statement issued in 1920 claimed that production had reached sixty tractors a week and that the target had been raised from 100 to 200 a week. By the middle of 1922, the company claimed a total of 3000 tractors had been built. Although the French market appears to have been satisfactory, demand in Britain was not, and production at the Birmingham factory ended in the mid-1920s.

Austin production in France, using engines brought in from Birmingham, continued. The French tractor was equipped with a three-speed gearbox, instead of the two speeds which were standard on the British version, and improvements were made to the specification, including the addition of a power take-off as an optional extra.

In 1930, the company decided to relaunch the improved French-built Austin in Britain. To help to re-establish the tractor, petrol and paraffin models were entered for the World Trials at Oxford, and both performed satisfactorily. Maximum belt power on the dynamometer was 24.1 hp for the petrol tractor and 19.1 hp with paraffin.

With tractor demand at a low level in Britain, and the Fordson still dominating the market, the relaunch was not a lasting commercial success. The Austin disappeared from the market in Britain after 1931, and disappeared from the French market soon afterwards.

McLAREN MOTOR WINDLASS

There were three companies in Britain with cable-ploughing systems powered by internal combustion engines, which were competing for the shrinking market during the early 1920s. All three were in the Leeds area of Yorkshire.

One of the companies was J. & H. McLaren. They announced a redesigned motor windlass in 1920 which was lighter, smaller and less expensive than competing equipment.

The McLaren system was also easy to operate, the company claimed, and was suitable for anyone who had 'a working knowledge of a motor bicycle or a small petrol engine'.

While Walsh and Clark, and the earlier Fowler cable engines, had followed the general styling of

McLaren's tractor and windlass for cable-ploughing.

traditional steam engines, the McLaren equipment was a complete break with traditional ideas of what a cable engine should look like. The

motor windlass consisted of a steel girder frame on four small wheels, carrying the engine and the vertical, rear-mounted winding drum. There was a simple chain and sprocket drive to the drum, with a choice of two working speeds.

A cable set consisted of two motor windlasses, 450 yards of cable, a seven-tine turning cultivator and a two-way balance plough to turn four furrows.

Operator comfort and safety were given scant attention when the McLaren engineers were designing the motor windlass. There was no canopy to provide protection from the weather, no seat to use when the windlass was operating with a cable, and the main driving chain and cogs were within easy reach of the operator's hands or clothing.

A four-cylinder petrol–paraffin engine provided power to drive the windlass and also to propel the unit from field to field. The power

output was claimed to be 40 hp on petrol and 32 hp when running on paraffin. This was sufficient to achieve a workrate of 1.43 acres an hour, which appears to be a fairly modest performance from a double-engine set with, presumably, a team of three operators.

The fuel consumption was 2.02 gallons of paraffin per acre ploughed. The manufacturers advised owners to reduce fuel consumption by stopping one of the engines while the opposite engine was winding in the cable. The engine could be restarted simply by engaging the drive to the drum while the last few yards of cable were still being pulled out. The tension on the cable would continue to turn the drum and also run the engine to start it again.

In spite of this energy-saving technique – which could not be used with steam equipment – the economics of cable-cultivation systems became increasingly difficult to justify. McLaren was still trying to sell cables sets in 1927 when the market had almost ceased to exist.

TWIN CITY 12–20

Twin City tractors were built by the Minneapolis Steel and Machinery Co. and were designed initially for the big acreage farms which could afford the cost of heavyweight equipment.

The company produced some of the biggest tractors on the American market until a change of policy, in about 1918, brought the first of the company's new smaller tractors into production.

These were the 16–30 tractor, followed in 1920 by the 12–20 (Plate 19), the smallest tractor the company produced. These were the forerunners of the range of lighter tractors which MSM built, until the 1929 merger to form the Minneapolis–Moline Power Implement Company.

Twin City.

MSM had originally designed the engine used in the 12–20 as the power unit for a truck and it was probably unique at that time as a tractor engine. It had four cylinders, each with twin inlet and exhaust valves operated by two overhead camshafts. This arrangement presumably meant a production-cost penalty compared with the simpler designs used by other manufacturers.

In its tractor version, the engine was adapted to run on paraffin and developed its rated power output at 1000 rpm.

The 12–20 engine probably helped Twin City to build up a reputation for advanced engineering features, which continued after the 1929 merger. When the merger came, the 12–20 tractor was no longer available, having been replaced by the more powerful 17–28, which was later adopted as part of the Minneapolis–Moline range.

INTERNATIONAL HARVESTER EXPERIMENTAL STEAM TRACTORS

The idea of using up-to-date technology to bring the steam engine back to the farm has attracted engineers on both sides of the Atlantic – and there are still some who believe that steam will one day make a come-back.

A small American company, the Bryan Harvester Co., manufactured modest numbers of steam tractors for about 5 years from 1922 (see *Great Tractors*) and did more than any other manufacturer to show how competitive a steam engine could be.

International Harvester was working along similar lines, producing at least two different experimental designs between 1920 and 1923.

This version of the International steam-tractor projects looks like a pre-production prototype.

An early version of the International Harvester
experimental steam tractor.

The steam tractor programme represented a substantial investment in time and money for IH but it never passed the experimental stage.

The first of the prototypes was simply a research vehicle, consisting of a chassis with various components attached to it. The rectangular shape over the front wheels consisted of a condenser with a large powered fan to reduce vapour from the pistons to water for recirculation.

The oval tank behind the condenser contained paraffin, which was the fuel used to raise steam in the boiler. The shiny black shape next to the fuel tank contained the boiler tubes, well insulated to conserve heat.

Steam from the boiler was fed to the two cylinders, which were in a heavily insulated V-formation over the rear axle. From the pistons, the steam was piped back to the condenser ready to repeat the cycle again.

This tractor was completed in 1921, and was followed by a redesigned version in 1922. The second tractor was more compact, with the condenser now resembling the radiator of a water-cooled engine and with some sheet metal-work to provide some front end styling.

A final version, based on the second tractor, had a finish which suggests that it may have been regarded as a pre-production prototype.

The production version never appeared. The tractors presumably shared the advantages of the Bryan, with a compact size, rapid warm-up to working pressure, accurate burner control and a closed water system which would need only occasional topping-up.

Exactly why International abandoned the project after taking it so far is not clear. They may have been disappointed by the performance of the experimental tractors they built, or perhaps the poor sales achieved by the Bryan tractor may have been discouraging.

Another theory is that the Farmall project, at an advanced stage of development in 1922, may have appeared more commercially attractive in a market which had beome highly competitive and unprofitable.

Another stage in the International Harvester steam-tractor project.

PETERBRO

The company which built the Peterbro (Plate 21) described it as 'the highest grade high-powered agricultural tractor ever built' in one of the sales leaflets they produced. The tractor was made at Peterborough by Peter Brotherhood Ltd, an engineering company with a long history in the railway industry.

Tractor production started in 1920 and the Peterbro was shown for the first time in the Royal Show of that year and also took part in the Royal Agricultural Society's tractor trials near Lincoln. In the class for tractors of 25 to 30 hp, the Peterbro took second place and a bronze medal, beaten by a British Wallis.

An unusual feature of the tractor was the engine, which was a Ricardo design of considerable complexity. The engine was designed to start on petrol and to run on paraffin. One of the characteristics of a paraffin-burning engine is unburned fuel finding its way past the piston rings and diluting the lubricating oil in the sump. Ricardo tried to overcome the problem by using a cross-head design and by the use of tiny bleed holes in the cylinder walls to divert liquid away from the sump.

Peter Brotherhood claimed that their engine design was unique and exclusive. It was also expensive and this helped to keep the sales volume small. Ironically, it was an engine failure which caused a Peterbro tractor to withdraw

(Right) An advertisement for the Peterbro.

from the World Trials near Oxford in 1930. A loose nut, it was thought, damaged a piston, connecting rod and splash guard.

The engine developed about 30 hp at 900 rpm and the pulling power was claimed to be 3750 lb at the drawbar in first gear. There were only two forward ratios, with a top speed of 3.5 mph in second.

The British market for tractors of more than average power and weight was small, but the Peterbro appears to have been more popular in Australia and New Zealand. The company concentrated on these two markets and the tractor won silver and gold medals at shows in New Zealand during the 1920s.

A new half-track version of the tractor was announced in 1928. The tracks were provided by Roadless. After that the next important event in the Peterbro's history was the 1930 World Trials and the engine misfortune in such public circumstances presumably did much to damage the tractor's commercial future.

(Below left) The Peterbro tractor from Peterborough.

(Below right) The engine used to power the Peterbro tractor.

THE PETERBRO' TRACTOR

MANUFACTURED BY

Peter Brotherhood, Limited, Peterborough.

**BRONZE
MEDAL,**
LINCOLN, 1920.

For use with Howard Ploughs, Cultivators, Knapp Drills, Garrett
Threshers and other Farm Implements.

AGRICULTURAL & GENERAL ENGINEERS, Ltd.

CENTRAL HOUSE, KINGSWAY, LONDON, W.C.2.

ASSOCIATING :

Aveling & Porter, Ltd. (Rochester).
Barford & Perkins, Ltd. (Peterborough).
E. H. Bentall & Co., Ltd. (Heybridge).
Blackstone & Co., Ltd. (Stamford).
Peter Brotherhood, Ltd. (Peterborough).
Charles Burrell & Sons, Ltd. (Thetford).
Burrell's Hiring Co., Ltd. (Thetford).

Clarke's Crank and Forge Co., Ltd.
  (Lincoln).
Davey, Paxman & Co., Ltd. (Colchester).
Richard Garrett & Sons, Ltd. (Leiston).
James & Fredk. Howard, Ltd. (Bedford).
L. R. Knapp & Co., Ltd. (Clanfield).
E. R. & F. Turner, Ltd. (Ipswich).
A. G. E. Electric Motors, Ltd. (Stow-
  market).

Front end view of the Peterbro.

It vanished from the market at some time in the early 1930s, at a time when the tractor industry was facing depressed markets. One of the few surviving Peterbro tractors is preserved at the Hunday National Tractor Museum in Northumberland.

SAUNDERSON LIGHT TRACTOR

'The design, the weight, the degree of efficiency, the materials, and even the price of £195 only are all outstanding features in a machine which, we make bold to say, will do much to restore the confidence of farmers in tractors.'

This was the way *Implement and Machinery Review*, the journal of the British farm equipment industry, welcomed the new Saunderson tractor in 1922.

The Saunderson company at Elstow, Bedford, had been Britain's biggest tractor manufacturer and the new model at the 1922 Smithfield Show attracted considerable interest. Its arrival was long overdue to meet the rapidly increasing competition from other new makes and models already on the market.

One of the unusual features of the new Saunderson was the engine design. This was a two-cylinder petrol–paraffin unit, with the cylinders arranged in a V-formation. According to the manufacturer, this design gave outstandingly smooth operation and the claim appears to have been supported at the demonstration held the following year at Stagsden, Bedford, when press reports were enthusiastic about the tractor's exceptionally smooth power unit.

The engine was water-cooled, and developed 20 hp at 1200 rpm, with 12 hp available at the drawbar. There was magneto ignition with an impulse starter and the manifold design was claimed to give especially rapid warm-up from cold to minimise the time required before switching from petrol to paraffin.

A cone clutch, which also incorporated one of the tractor's two braking systems, transmitted the drive to the gearbox, with two forward ratios and a reverse, giving the tractor a maximum forward speed of 3 mph. The drive to the differential, where the second braking system operated on the main shaft, was by means of a roller chain. The final drive to the rear wheels was also by chain and sprocket. The drive chains were all fully enclosed.

One of the special features of the new tractor was its light weight and this was emphasised in a substantial advertising campaign. Soil compaction problems were familiar in arable farming

Saunderson's last production model.

A Saunderson at work with a threshing drum.

areas at that time and tractors were blamed for much of the damage. The Saunderson's 27 cwt was considered reasonably safe.

Another sales feature was the Saunderson warranty. This provided protection for a remarkably generous 3 years, giving the owner a limit on the cost of replacement parts each year. For the first year, the maximum the owner would spend was £5, and any additional cost for parts was paid by the manufacturer. In the second year this limit was £10, increasing to £15 in the third year. This was designed to make the operating cost of the tractor more predictable, as it included parts to be replaced through normal wear and tear.

Unfortunately, the warranty probably proved to be of doubtful value. In 1924, H. P. Saunderson decided to sell his ailing tractor company to the Crossley engineering group of Manchester and tractor production ended soon afterwards.

Under those circumstances, the unexpired portion of Saunderson warranties might not be accepted by the new owners.

Plate 16 Rumely Oil Pull 12-20 model.

Plate 17 Rumely Model R 25-45.

Plate 18 A British-built Austin tractor.

Plate 19 Twin City 12-20.

Plate 20 Heider 15-27 at the Manitoba Agricultural Museum.

Plate 21 Peterbro tractor.

Plate 22 Hart-Parr 28-50

Plate 23 International Harvester Farmall F12 with mounted potato-spinner.

Plate 24 International Harvester Farmall F14.

Plate 25 A Cassani tractor on display at the SAME offices in Treviglio, Italy.

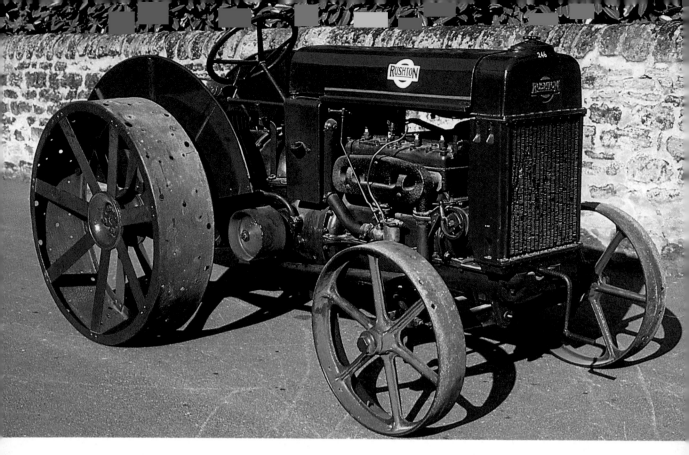

Plate 26 Rushton tractor.

Plate 27 Renault PE tractor, believed to be the only one in Britain.

Plate 28 Case Model R.

Plate 29 Massey-Harris Pacemaker tractor.

Plate 30 Caterpillar R2 tracklayer.

Plate 31 The Oliver 28-44 in the Manitoba Agricultural Museum.

JOHN DEERE MODEL D

Deere and Company bought their way into the tractor market when they took over the Waterloo Gasoline Engine Co. in 1918. This gave them the Waterloo Boy tractor to sell (see *Great Tractors*).

Although the Waterloo Boy was well known and had a good reputation, the design was already becoming outdated when the tractor was taken into the Deere product line. This meant that a more modern replacement model was needed to give the company an opportunity to expand its share of the market.

The new tractor arrived in 1923. It was the Model D, the first tractor to carry the John Deere name and the beginning of a production run which continued, with modifications which included restyling, until 1953.

During this period, the horsepower recorded in Nebraska tests increased from 27 hp on the belt pulley in 1924 to 38.11 hp when the last Model D

(Overleaf) John Deere Model D tractor built in about 1926.

(Below) An early 1930s' version of the John Deere Model D.

test was carried out in 1940. In the same tests the drawbar power rose from 15 to 30.46 hp, which is an improvement in efficiency from 55 per cent of the belt hp to 80 per cent in 16 years.

The Waterloo Boy engine was a two-cylinder unit and the same design was chosen for the Model D. With one exception, the 1937 Model L, every John Deere tractor until 1960 was powered by two horizontal cylinders.

While other large-scale tractor manufacturers were moving away from two cylinders, John Deere were making a considerable success with their design.

Farmers liked its simplicity, the company claims, with fewer parts to wear or repair and easier servicing. Even the cooling system contributed to this, with a thermosyphonic design instead of a water pump. The engines were sturdy enough to give a good standard of reliability, operating on various low grade fuels.

Mechanical simplicity included the transmission. There were only two forward ratios and the clutch was operated by means of a hand lever. The final drive to the back axle was by a chain and sprocket. A power take-off kit was available as an option in 1925 and the design of the pulley was changed in 1931, providing a guide to identifying the earlier tractors built during the 1920s.

HEIDER 15–27

The first Heider tractors were built in 1911 at Carroll, Iowa. One of the unusual design features was the engine position, which was directly over the rear axle. This presumably helped to encourage pulling efficiency, with the weight of the engine to reduce wheelslip.

Pulling power may have been one of the factors contributing to the success of the tractors. Their popularity received a further boost when the manufacturers made an agreement for the Rock Island Plow Co. of Rock Island, Illinois, to share the marketing.

Another deal between the two companies followed in 1916. This time, the Rock Island company bought the manufacturing rights to the Heider design. Heider pulled out of the tractor business to concentrate on making implements, while Rock Island expanded production.

The deal which transferred the tractors to the Rock Island Plow Co. also included the right for the new owners to continue to use the name Heider. They continued to do so with considerable prominence for a further 10 years or so.

Under the new ownership, the tractor range was expanded and Heider became one of the more familiar makes in the medium-size sector of the American market.

The 15–27 (Plate 20) arrived in 1924. The roof over the rear of the tractor carried the decorative fringe which had become a distinctive feature of Heider tractors since before the Rock Island takeover.

Like earlier Heiders, the new tractor was equipped with a Waukesha engine operating on paraffin. This was a four-cylinder vertical design, with 4.75 in bore and 6.75 in stroke. At Nebraska in 1925, it produced 27.16 hp at 900 rpm in the belt test.

At that time, the 15–27 was the most powerful model in the range and it remained in production until 1927, when a more powerful tractor arrived. The newcomer, listed as an 18–35, was marketed under the Rock Island name, and this was the name which was used until the company was taken over by Case in 1937.

The 15–27 was the last new model to carry the name Heider and the last with a fancy fringe around the driving area roof.

HART–PARR 12–24 AND 28–50

When Charles Hart and Charles Parr completed their first tractor in 1902, it was powered by a two-cylinder horizontal engine. This remained almost the standard type of engine for Hart–Parr tractors until the company lost its identity as part of the Oliver organisation in 1929.

There were a few exceptions to the rule, with a very small number of single- and four-cylinder models and some vertical engines. Sometimes, when a four-cylinder engine was specified, it was simply two of the twin-cylinder power units placed side by side.

One example of this doubling-up technique to provide more power was the 28–50 tractor, which was available from 1927 until the new range of Oliver tractors arrived in 1930. The 28–50 was the most powerful model in the Hart–Parr range at the time and the makers claimed it could handle a six-bottom plough or a 36 in threshing drum.

A Hart–Parr 12–24 tractor, beautifully restored at the Henry Ford Museum and Greenfield Village Collection at Dearborn.

The smallest model in Hart–Parr's late 1920s' range was the 12–24, rated as a three-bottom plough tractor and suitable for a 24 in thresher.

Hart–Parr launched the 12–24 in an E version in 1924, powered by twin-cylinder engine with horizontal cylinders of 5.5 in bore and 6.5 in stroke. This was replaced in 1928 by the H version, which was on the market until 1930.

The E and H versions of the 12–24 had basically the same engine design, but with a slight increase in the cylinder bore for the later model. It was the bigger capacity engine which was doubled-up to provide the four-cylinder power for the 28–50.

In the big tractor, the two engines were squeezed in side-by-side, but operated at the same rated speed of 850 rpm used in the 12–24.

The twin-engined 28–50 (Plate 22) was almost twice as heavy as the 12–24 and, at $2085, was almost twice as expensive as the $1050 12–24.

At Nebraska, the two engines of the 28–50 achieved 64.56 hp in the maximum load belt test, against 31.99 hp for the single-engined tractor. Both tractors easily exceeded their drawbar rating in the tests, with a maximum load pull of 45.58 hp and 21.78 hp respectively. Fuel consumption related to power output was similar for both tractors. The 12–24 and 28–50 tractors were among the last of the Hart–Parr line, with almost 30 years of twin-cylinder power.

INTERNATIONAL HARVESTER FARMALL

With the arrival of the Farmall, the tractor had at last become the year-round power unit which could completely replace the horse. It happened in 1922, more than 30 years after the development of the first tractors.

The man who developed the Farmall was Bert R. Benjamin, a senior development engineer at International Harvester's Chicago headquarters. The company's records suggest that he had started working on ideas for a really versatile rowcrop tractor in 1915, and that his first prototype was a motor-cultivator design which lacked the pulling power to achieve a satisfactory ploughing performance.

Benjamin's basic ideas for what later became the Farmall were completed in the early 1920s and the company decided to give the project high priority in an attempt to revive sales in a market which was becoming dominated by Henry Ford's price-cutting policy.

A pre-production batch of Farmalls was completed in 1922 for field testing. Most of these were sent to farms in Texas to be used in both cotton and corn. The tractor went into production in 1924, with an original price of $950, plus an additional $88.50 for the specially-designed mounted cultivator.

The original design, which remained in production with various modifications and different engines until 1939, had a tricycle wheel arrangement with twin front wheels, popular for cultivation work in the USA, but less suitable for the heavier, wetter soil conditions of Europe.

Among the features which encouraged some farmers to buy a Farmall, instead of the very much less expensive Fordson, was the high rear axle clearance for straddling rows. This was achieved by having the final geared reduction for the rear wheels housed at the outer ends of the half shafts.

Another design feature was the independent brake system, which could lock either of the rear wheels to achieve a pivot turn in a confined space. The brake system could be linked to the steering mechanism so that the pivoting became automatic.

Implements could be towed conventionally, or mounted at the front of the tractor on a frame, or mid-mounted for really accurate inter-row working.

The four-cylinder engine developed 18 hp on the belt in its Nebraska test and the tractor comfortably exceeded its 9 hp drawbar rating. The tractor recorded a weight of 3350 lb on the

International Harvester Farmall F14.

Nebraska weighbridge and used an exceedingly modest 1.87 gallons of fuel an hour in the rated load belt test.

Demand for the Farmall increased rapidly and its success attracted plenty of imitation as a new generation of rowcrop tractors arrived on the market. The Farmall did well to hold the lead in the new market it had created and, in 1931, the original model was replaced by the new Farmall F20 and F30 tractors.

The F20 was the direct replacement for the previous Farmall, powered by an uprated version of the earlier engine with a 10 per cent increase in horsepower. The F20 remained in production until 1939, when the original Farmall design was discontinued.

The F30 was simply a bigger Farmall with a 32 hp engine and a three-plough classification. The engine was a modified version of the power unit used in the 10–20, and the sales literature claimed there was sufficient power to plough up to 13.5 acres a day or to operate a 28 in thresher.

Further development of the Farmall range came in 1932, with the launch of the new F12 (Plate 23), with a petrol engine rated at 16 hp. A petrol–paraffin version arrived later and, in 1938, the power was raised to convert the one-plough F12 into the two-plough F14 (Plate 24).

During its long and commercially successful production life, the Farmall retained all the original features, including the distinctive steering linkage with a long horizontal rod over the engine, linked to a vertical column over the front axle. But there was also an increasing list of optional equipment, ranging from a cushioned seat for the driver to a big choice of implements.

There was a full width front axle, as an alternative to the dual wheels of the original tricycle design, and a long list of rear-wheel designs to suit various soils and row widths, with a rubber-tyre option arriving in 1934.

During the 1920s and early 1930s, when the tractor market was often in difficulties, many manufacturers pulled out of the market and few of those who remained could claim a real success story. The Farmall was a success and a genuine step forward in the development of power farming.

RONALDSON–TIPPETT

Ronaldson Bros and Tippett, for many years the largest manufacturer of engines in Australia, was founded at Ballarat, in Victoria, by David Ronaldson in 1903.

Production started with a simple range of barn machinery, including chaffcutters and grain crushers, but stationary engines were added to the product line in about 1905.

Demand for engine power was increasing and David Ronaldson's company expanded rapidly. They built a wide range of power units, including marine diesels and engines for power stations and for portable generating sets.

David Ronaldson designed the company's first prototype tractor in about 1912. The power unit was a Ronaldson-Tippett engine with a kerosene vaporiser. The engine was water-cooled, but details of its size and power output are no longer available.

(Right) A batch of Ronaldson–Tippett tractors at the factory awaiting despatch to customers.

(Inset) A batch of the first production version of the Ronaldson–Tippett, nearing completion in the factory.

(Below) This was the first Ronaldson–Tippett tractor. It never went into production.

The tractor proved to be unsatisfactory, apparently because it was too heavy for the power available, and it was eventually scrapped.

A second attempt to move into the tractor market came in 1926 and this was much more successful. The tractor was rated at 30 hp, and was designed to suit Australian conditions. It sold in substantial numbers against imported competition until production ended in 1938.

Engines for the new tractor were bought in from the Wisconsin company in the USA. Ronaldson Bros and Tippett had no suitable engine of their own and, presumably, decided it would not be worthwhile to develop one especially for the tractor.

The engine was a four-cylinder unit with the cylinders in two separately water-jacketed pairs. The original cooling system proved inadequate for the high temperatures which occur in

(Top) Ronaldson–Tippett tractor at work.

(Above) Later version of the Ronaldson–Tippett showing the increased cooling capacity.

Ronaldson–Tippett tractor showing the engine.

Australia. To cure the problem, additional water capacity was provided by means of an extension above the radiator. The design change appears to have been introduced after about a hundred tractors had been built and gives the later version a distinctive appearance.

The transmission system had several unconventional features. There was a hand lever to operate the clutch and this also controlled the throttle. As the lever was pushed forward to engage the clutch, the throttle opening increased to provide more power. Pulling the lever back towards the driver disengaged the drive and brought the engine back to idling speed. There was also a separate lever to preset the throttle opening.

Only two forward ratios and a reverse were provided inside the gearbox, but there was also a pair of external pick-off gears which could be interchanged by hand to provide alternative travel speeds from the two internal ratios. To increase the permutations further, there were alternative pairs of external gears available, and a choice of 50– or 53 in diameter driving wheels.

According to the comprehensively-detailed instruction book, the two wheel sizes, six sets of pick-off gears and the high and low ratios inside the gearbox provided a choice of forty-eight forward speeds at maximum throttle setting. This was doubled to ninety-six by providing full sets of figures for both half and full penetration of the 5 in wheel lugs. All of these speeds were in a range from 1.8 to 6.6 mph.

The footbrake operated on a drum attached to the pulley shaft and was effective only when the tractor was in gear. The instruction book included a warning about the perils of disengaging the drive in order to change gear on a steep hill. It

A photograph of the radiator, engine and gearbox assembly taken originally for use in the Ronaldson–Tippett instruction book.

(Inset) Another photograph showing how pick-off gears could be interchanged to provide a different range of speeds.

was essential to put blocks behind the wheels to prevent the tractor rolling down the hill with the brake uselessly locking the pulley drive.

Droughts and the financial depression which hit Australian farming in the early 1930s caused serious problems for tractor salesmen. Ronaldson Bros and Tippett were forced to repossess a large number of the tractors they had already sold. These had to be reconditioned and were often sold again at a considerable loss.

The situation occurred at a time when the company should have been developing a new model with improvements to match the more up-to-date imported models which were available. While the company's engine production survived, and expanded rapidly during World War 2, the Ronaldson-Tippet tractor disappeared from the market.

CASSANI

Francesco Cassani was one of the pioneers of the Italian tractor industry. The company he established now builds SAME and Lamborghini tractors and also controls the Hurlimann factory in Switzerland.

Cassani's special interest was diesel engine development and he designed the first high-speed marine diesels used by the Italian Navy. He also designed several engines for the Italian

Cassani tractor on display at the company's headquarters.

A photograph taken during a demonstration in 1927 or 1928.

Air Force, including an 800 hp diesel with eight cylinders, which was completed in prototype form as the Second World War was coming to an end in 1944.

His first tractor was completed in 1927, powered by a two-cylinder diesel engine of his own design. At that time, Cassani was only 21 years old and all his engineering skills had been picked up in his father's workshop, where farm implements were made and repaired, and at evening classes.

The 1927 tractor was probably the first in Italy with a diesel engine and was a year ahead of the British McLaren diesel tractor and 4 years in advance of the first Caterpillar diesel in the USA. The Cassani engine was water-cooled with horizontal cylinders.

The prototype tractor was followed by a pro-duction version (Plate 25) with an engine rated at 40 hp. This used a compressed-air system for starting the engine, with an air tank over the engine. While the tractor was working, air was pumped into the tank until the required pressure was reached.

Italian farmers have a traditional liking for very deep ploughing and this presumably ex-plains the relatively high power of the Cassani tractor. It appears to have been popular in Italy during the early 1930s and several examples have survived.

Later, the tractor market ran into a recession and Cassani concentrated on his engine develop-ment work, which was mainly for the Italian Government at that time. One of his successes was a new type of pump for high-speed diesels, which was patented and manufactured in large numbers.

Cassani returned to tractor development in 1942, and formed a new company under the

SAME name. The SAME group of companies now manufactures both air-and water-cooled engines, and claims the fourth largest share of the European tractor market.

One of the early versions of the Cassani, now preserved at the SAME factory in Italy.

RUSHTON

Several companies attempted to produce new tractors to sell in direct competition with the Fordson. Presumably this seemed a good idea at the time, but the price and reputation of the Fordson F and N models at the peak of their popularity put many competitors out of business.

George Rushton realised that he could not beat Henry Ford on price, but he was confident he could produce a tractor with some extra features which would sell in large numbers for a little more money than the Fordson.

Rushton worked for the AEC concern, famous for the double-decker buses supplied to the London General Omnibus Co. His tractor project was given substantial support by AEC and the tractors were built at an AEC factory at Walthamstow, London.

When the tractor project was first announced to the public in 1928, the trade name 'General' was used and this was the name on the front of the

Rushton tractor with Fordson-type engine.

tractors, in the logo which was familiar on thousands of London buses. A further announcement in the following year said that a new company had been formed to operate the tractor business and the trade name was changed to Rushton.

The company appears to have gone out of its way to stress that the tractors had been introduced to compete directly against the Fordson. The tractors were claimed to have all the good features of the Fordson, plus some additional features of their own. The 1928 announcement also mentioned that E. Allan Webb had given up his position as Tractor Manager for Ford in Britain to become sales director for its new British-built rival.

There was an even closer link between the Rushton and the Fordson. Many of the Rushton components were so similar to those used in the Fordson that they were interchangeable. It has been suggested that the first few General and Rushton tractors were built with a considerable proportion of Fordson parts, while the supply of British sourced items for the Rushton factory was being organised.

Extra features which George Rushton hoped would justify the higher cost of his tractor included extra radiator capacity, a self-cleaning system on the rear wheels, magneto ignition (while Fordsons were still using a coil) and a little more power through a slightly increased engine capacity.

The two tractors competed against each other at the World Trials, which were held at Oxford in 1930, with entries from eight countries. There were two Rushtons, one a standard petrol–paraffin model (Plate 26), and the other with a higher compression ratio to operate on petrol and equipped with Roadless tracks. Unfortunately, a full comparison of results is not possible because the Fordson was withdrawn with a cracked cylinder block, but where test data are available for both makes, the Rushton showed some advantages.

For the maximum load test on the belt, the standard Rushton recorded 23.9 hp against 20.8 hp for the Fordson. The Roadless version of the Rushton produced 32.2. hp on petrol. The Fordson achieved a slightly better fuel-efficiency figure, but also used more cooling water.

Although Rushton tractors appeared to perform quite well, and were strongly marketed in Britain and overseas, there were never enough customers. George Rushton's venture had been launched at a difficult time for the tractor industry, with too many companies competing for a very limited number of sales and financial clouds causing increasing concern around the world.

The situation deteriorated in the early 1930s and AEC financial support was apparently no longer available. George Rushton left the company, which continued to operate on a small scale. There was more bad news when the production of Fordson tractors was transferred from Ireland to the new factory at Dagenham, Essex, not far from Rushton's Walthamstow plant.

The end of the Rushton story came in the mid-1930s, leaving the Fordson as easily the top-selling British tractor.

RENAULT PE

Renault's history as a tractor manufacturer goes back to the GP model of 1919, which had been developed from experience in designing a light tank during the war (see *Great Tractors*). But by then the company was already some 20 years old and had already established an international reputation in the car industry.

The Renault association with vehicles began in 1898 when Louis Renault built a small car, allegedly in his parents' garden shed. More cars followed, home-made at first but, soon made on a commercial basis. Renault entered his cars in many important sporting events which helped to publicise them in the period before World War 1.

Renault's GP tracklayer and its wheeled derivatives established the company as France's biggest tractor manufacturer during the 1920s. They were replaced by the PE model (Plate 27), which was under development from 1927 and arrived on the market in 1929.

When the PE tractor was on the drawing board, the designers retained the traditional Renault arrangement of locating the engine in front of the radiator. This had been a feature of the previous tractor and had also been a characteristic of most Renault cars. It allowed the stylists to give Renault vehicles a low profile at the front end, providing good forward visibility and, presumbably, a fairly low drag coefficient.

Although the Renault continued the back-to-front engine and radiator positions, the front end of the tractor looked quite ordinary. This was because an enormous air cleaner filled the space which would normally be occupied by a radiator. The air cleaner was filled with a dry element which must have had an exceptionally large capacity for trapping dust.

The engine was a four-cylinder, in-line design, petrol-operated with 75 mm bore and 125 mm stroke. It was cooled by a rather complicated arrangement, dictated by the radiator position.

(Top) Renault PE tractor.

(Above) This picture shows the Renault engine located in front of the radiator.

(Overleaf) An early photograph of a Renault PE tractor.

An obvious way to cool the engine would be to put a fan against the radiator to draw air into the

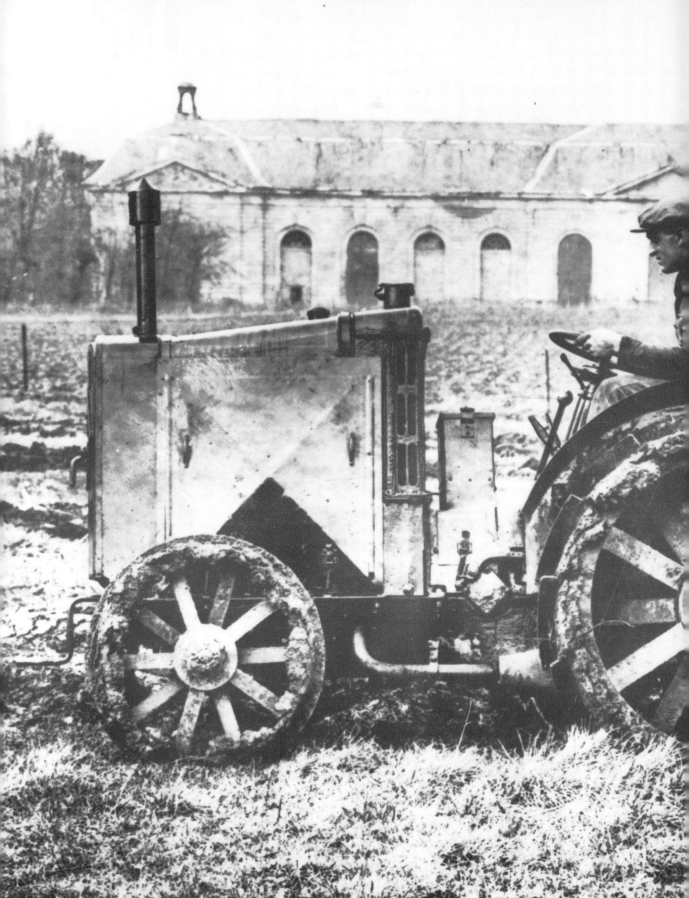

engine compartment through the cooling fins. The designers chose not to arrange it in this way!

Instead, they designed a fan which was part of the clutch, positioned under the bottom of the radiator. This was designed to draw air out of the engine compartment, to be released beneath the middle of the tractor. The cooling effect was achieved by cold air passing through the radiator to replace the air which the fan had removed. To ensure that the air was drawn through the radiator, pressed steel panels provided close-fitting sides for the engine compartment, to force the incoming air to travel the required route.

The PE, which remained in production for about 4 years, was the last Renault tractor to use this cooling arrangement. The 1933 replacement model carried its radiator at the front and survived with a standard-size air cleaner.

FERGUSON BLACK TRACTOR

The introduction of the Ferguson System of implement attachment and control was one of the most important developments in the history of power farming. It was used for the first time in 1933 on the Black Tractor, a prototype built by Ferguson and his team to demonstrate the new system of three-point linkage and hydraulic control which they had developed.

Harry Ferguson's ideas for improving the attachment of implements to tractors originated during World War 1, when he was still living in what is now Northern Ireland. His garage business in Belfast took the agency for Waterloo Boy tractors, which were sold in Britain as the Overtime.

He took a close personal interest in demonstrations to sell tractors to local farmers and he also visited farms throughout Ireland for the Board of Agriculture, helping to improve the standard of tractor operation for the wartime ploughing campaign.

It was this experience with tractors and implements of various makes which persuaded Harry Ferguson to look for a more efficient way to use tractor power. He started working in 1917 on what was eventually to become the Ferguson System, helped by William Sands who was employed in the Ferguson garage. The first practical result of their work was a mounted plough designed for use with the Eros tractor conversion for the Model T Ford car. It was followed by a second version to fit the Fordson tractor.

Although the Ferguson ploughs were fully mounted, with a spring-assisted lever for raising them out of work, they were still a long way from the complete Ferguson System. This was developed gradually with the addition of hydraulic draft control and three linkage points.

The decision to build a special tractor to demonstrate the system came in 1932. The implement attachment system was by this time complete, although further refinements came later and Ferguson wanted to bring it onto the market. Although various companies had expressed some interest in his ideas, none had made a firm commitment.

The problem of finding a manufacturer was hardly surprising. At a time when tractor companies were going out of business in the USA and Britain, few industrialists were willing to make a large investment in a completely new tractor requiring a special range of matched implements.

Harry Ferguson decided that the best way to convince people that the new system he had developed was worth backing, was by demonstrating how well it worked. In order to be able to do this, he had to have a Ferguson System tractor.

His Black Tractor was designed and built during 1932–3 with the care and thoroughness which were characteristic of everything he did in his business life. The tractor incorporated not only the essentials of the implement attachment

Harry Ferguson.

system, but all of his other very firmly held ideas about tractor design. The Black Tractor was small, light and beautifully finished in every detail.

The glossy black paint finish was probably chosen to help emphasise the functional neatness of the design. Paint colour was the sort of detail which Harry Ferguson considered with care. All the production tractors with which he was closely involved were painted battleship grey, until he lost control over such matters after the Massey–Harris takeover.

Components for the tractor were either hand-made or bought-in. The engine was supplied by the Hercules company in the USA, and was a four-cylinder unit producing about 18 hp. Gears were supplied by the David Brown company of Huddersfield, together with some components for the steering mechanism.

When the tractor was complete in 1933, it was tested with various implements which had been specially designed and built to demonstrate the system. One important design change resulting from the tests was to use compression loads on the top link to operate the hydraulic control valve, instead of tension loads in the lower link arms as had been originally arranged.

Henry Ford had established a tractor factory in Cork, because he wanted to make a practical contribution to the economy of the region in Ireland from which his father had emigrated to the USA. Harry Ferguson hoped that the first Ferguson System tractors could be built in Northern Ireland, where he was born and brought up, and the Black Tractor was taken there for its first demonstrations.

Harry Ferguson's idea of a tractor factory to provide additional employment in Northern Ireland failed to materialise, but later demonstrations with the Black Tractor back in England attracted more positive interest. They resulted in a partnership between Ferguson and David Brown, later Sir David, of the company which had supplied gears for the Black Tractor. Ferguson designed a production version of the

The Ferguson Black Tractor.

Harry Ferguson's Black Tractor.

Black Tractor and set up a marketing organisation to sell it with a range of specially designed
implements. David Brown looked after the
production.

The new tractors carried the Ferguson name,
but were often referred to as Ferguson-Brown
tractors (see *Great Tractors*). They were built as a
direct result of the successful performance of the
Black Tractor and they brought the Ferguson
System on to farms for the first time.

CASE MODEL R

Whoever was responsible for the styling of Case tractors during the 1920s and 1930s appears to have done a good job. Few tractors from that period attract as many favourable comments now, or have stood the test of time as well. Case L and V tractors are both good examples and so is the neat styling of the Model R.

The R was introduced in 1936 as the small model in the Case range and was available in several versions to suit specialised requirements. The standard four-wheel tractor was the Model R (Plate 28), with an industrial version called the RI, and the rowcrop model was the RC with a tricycle wheel arrangement.

Case have almost always used their own engines, but the Model R was an exception, with engines bought in from Waukesha. This was a petrol engine, with four cylinders of 3.25 in bore and 4 in stroke, rated at 1400 rpm.

The Model R and the rowcrop RC were tested at Nebraska. Power outputs recorded in the belt and drawbar tests show some interesting comparisons.

Case submitted the Model R tractor with rubber tyres fitted. This tractor registered about 20 hp in the maximum load belt tests and up to 18.28 hp on the drawbar. This shows an unusually small power loss through the transmission and driving wheels and suggests that the Model R on rubber tyres would be an efficient pulling tractor.

The RC was supplied to Nebraska with cleats instead of rubber tyres. The engine power registered in the belt tests was close to the figures achieved for the R, as would be expected, but the best result in the drawbar tests was only 14.21 hp, which is one of several examples produced at Nebraska of the advantage of using rubber for wheel grip.

Case dropped the Model R in 1940, replacing it with the V series, offering slightly more power and another example of good styling.

Case Model R, showing the distinctive styling of the radiator grille.

MASSEY–HARRIS PACEMAKER

One of the big events in the tractor industry in 1928 was the takeover of the J. I. Case Plow Works Company by Massey–Harris.

Behind the deal was the Massey–Harris company's determination to buy its way into the tractor industry. Two previous attempts, involving the Bull company in 1917 and the Parrett tractor in 1918, had both failed, leaving the Canadian company with a successful range of machinery, but without a tractor.

The Case Plow Works built the Wallis tractor in their factory at Racine, Wisconsin. This was a well-established tractor with a good reputation. Negotiations began in 1926 to arrange a marketing agreement, but ended with the takeover 2 years later.

A distinctive feature of the Wallis was the curved steel plate under the engine and gearbox. This was the unit frame originally developed in 1913 for the Wallis Cub and retained as a design feature in subsequent models.

Massey–Harris retained the curved underside as they introduced developments of the Wallis and it was still used when the new Challenger and Pacemaker models went into production in 1936.

The Pacemaker (Plate 29) was rated as a three-plough tractor in the USA. It was powered by a four-cylinder Massey–Harris engine which developed almost 30 hp at 1200 rpm in the original version of the tractor. An alternative version, known as the Twin Power Pacemaker, was announced in 1938 in which the engine speed could be increased to 1400 rpm, raising the power output to 42 hp.

The Pacemaker and Challenger models, which both used the same engine design, were the last tractors to carry the distinctive Wallis styling. The replacement was the 101 series, the first version of which arrived in 1938 and remained in production throughout World War 2.

A Massey–Harris Pacemaker photographed on a Canadian farm in the 1930s.

JOHN DEERE MODELS L AND LA

In 1930, there were more than 19 million horses and mules at work on American farms. Most of them were on the smaller farms where tractor power was still considered an expensive luxury.

For the tractor industry, these farms represented a big potential market and, during the 1930s, several manufacturers introduced new models aimed directly at the small-farm market.

John Deere's mini tractor was the Model L which arrived on the market in 1937. It was announced as 'The lightweight economical tractor that handles all work ordinarily done with a team of horses'. The sales leaflet claimed that the tractor could handle loads which would normally require a two-horse team, but pointed out that the tractor would do more work in a day because of its faster speed.

The power unit was a water-cooled twin-cylinder engine which developed 9.27 hp in the Nebraska belt test, with 7 hp at the drawbar. The engine was supplied by the Hercules company and operated on petrol. It developed its rated power at 1550 rpm.

Three forward gears provided a maximum travel speed of 6 mph with the standard governor setting. The speed could be raised to about 12 mph by using a special kit to adjust the governed engine speed to 2400 rpm.

With a turning circle of 7 ft and an overall length of only 7 ft 6 in, the Model L was suitable for working in small fields and confined spaces. For rowcrop work, the rear wheel tread could be adjusted between 36 in and 42 in, and this could be increased to 54 in by reversing the rear wheels. There was adequate clearance under the tractor for working with mid-mounted equipment and the forward visibility for precision work was improved by the slightly offset position of the engine.

As the Model L offered less power than many modern lawn tractors, the workrate for jobs such as ploughing must have been slow and, in some conditions, almost impossible. Even so there was sufficient demand for the tractor to keep it in the product line until 1946, and to add a slightly more powerful version to the range in 1941.

A John Deere Model L tractor powering a saw.

Clearing snow with a John Deere Model L.

The 1941 model was the LA, outwardly similar in appearance, but with slightly increased engine capacity to produce 12.93 hp on the belt and 10.46 hp at the drawbar. Other detailed differences included larger rear wheels and a faster engine speed. The special kit for raising the engine speed to make the tractor faster was not available for the LA.

The LA was available until 1946 and outsold the L during the 5 years that both were on the market.

CATERPILLAR D2 AND R2

Caterpillar was the first American company to put a diesel-powered tractor on the market and, during the 1930s, additional models in the company's range were equipped with diesel engines.

One of these was the D2, a small tracklayer introduced in 1938. The R2 Caterpillar (Plate 30) was basically similar, but with the choice of petrol or paraffin engines.

The D2 and both versions of the R2 were put through the Nebraska test programme in 1939. The tests were carried out consecutively and the data on performance provides an interesting comparison of the three fuels. All three engines were built by Caterpillar and were of the same basic design, with the same cylinder dimensions and cubic capacity.

During the belt tests, where engine differences are most apparent, all three engines were operated within 1 rpm of the rated 1525 rpm.

In the 1-hour maximum load test, the diesel engine produced 29.98 hp on the belt, compared with 28.95 hp for the petrol version and 27.78 hp with paraffin.

Larger differences showed up in the fuel consumption figures, which were 2.26 gallons per hour for the diesel, 3.56 gallons from the petrol engine and 3.09 gallons on paraffin. The most important figures for making fuel efficiency comparisons are those showing the hp-hours per gallon of fuel used, and this is where the advantage of the diesel is most apparent.

Each gallon of diesel produced 13.24 hp-hours against 8.62 hp-hours from a gallon of petrol and 8.99 hp-hours from paraffin.

Obviously there are other factors to assess in order to make a complete comparison of the three tractor versions. The D2 was more expensive, and also heavier than the R2. Part of the difference was the small petrol engine provided on Caterpillar diesels for starting the main engine.

The starter engine was a two-cylinder horizontal unit, which was started by pulling a cord. The capacity was 587 cc. The function of the small petrol engine was to turn over the four-cylinder diesel motor to start it and also to help warm up the water in the cooling system to encourage the diesel to fire more readily.

Caterpillar R2 tractor ploughing.

Caterpillar R2 followed by a Field Marshall at a vintage tractor event in England.

The Nebraska test reports also show that the advantages of the diesel engine were beginning to influence more tractor manufacturers in the late 1930s. Of nine different tractors tested in 1937, none had a diesel engine. There was one diesel, a Caterpillar D8, among the seventeen tractor models tested in 1938, four out of twenty-three in 1939, rising to nine among the twenty-seven tractors in 1940.

OLIVER 28–44

The 28–44 was one of a new range of tractors introduced in 1930 by the newly formed Oliver Farm Equipment Corporation of Chicago.

Four important companies had joined forces in 1929 to form the new corporation. These included the Hart–Parr company of Charles City, Iowa, and another manufacturer, the Nichols and Shepard Co. of Battle Creek, Michigan.

The name 'Oliver' originated with a Scottish immigrant, James Oliver, who began making ploughs in 1855 and formed the company which later became known as the Oliver Chilled Plow Co.

Tractor production for the new corporation was concentrated at the former Hart–Parr factory and the name Hart–Parr was retained until the mid-1930s.

One of the changes introduced with the new 1930 models was a more modern engine design. Hart–Parr had remained faithful to horizontal engines, which were almost always of two-cylinder type. These were replaced by four-cylinder vertical power units.

For the 28–44 (Plate 31), the most powerful model in the range, Oliver used a valve-in-head engine with 4.75 in bore and 6.25 in stroke, which developed its rated power at 1125 rpm. It was available in petrol or paraffin versions and used a high tension magneto.

One of the various versions based on the 28–44 model was an industrial tractor, which was the first Oliver with inflatable tyres. The engine was also sold as a skid unit for use in various industrial products by outside manufacturers.

Hart–Parr tractors had always been slow on the road and this characteristic was retained in the 28–44. There were three forward ratios, with a maximum travel speed of 4.3 mph. This was raised in the industrial model to 7 mph.

Oliver kept the 28–44 in production until 1937, when it was replaced by the 90 Series. This retained the old 28–44 engine, but added a self-starter. The three-speed gearbox was replaced by a new four-speed transmission, raising the maximum forward speed to 5.5 mph. The Oliver 90 remained in the price list until 1953 and the company which made it survived until 1960, when it became a subsidiary of White Motors.

FOWLER FD2

The FD2 tracklayer was part of an ambitious plan to revive the fortunes of John Fowler and Co. of Leeds.

During the great days of steam, Fowler had been one of the most successful companies, building up an international reputation for engineering quality and innovation. As the agricultural market for steam power faded, the company faced serious difficulties and the search for alternative products failed to halt the deteriorating financial situation.

The outbreak of war in 1939 filled the factory again as Fowlers became a leading manufacturer of tanks for the Allied armies. The British Government, apparently anxious to exercise more control over the company's management,

A Fowler FD2 photographed at a demonstration in South Africa.

made a compulsory takeover in 1941. This situation was temporary and the assets were sold in 1945 to Rotary Hoes, the Essex company which was introducing a range of rotary cultivators.

One of the assets acquired by Rotary Hoes was a range of diesel engines. These had been designed by Freeman Sanders, who joined Fowler in 1934 and became engineering director in 1936. Fowler had taken an early interest in diesel power and Sanders introduced some important improvements, which were patented in 1935. These included a two-way swirl combustion chamber design, allowing more efficient combustion with smoother running, cleaner exhaust and easier starting.

In 1944, with the end of the wartime tank contracts anticipated, plans were made to establish Fowler as a specialist manufacturer of crawler tractors. The strategy made sense.

Fowler FD2 tractor with its Rotavator attachment.

Suitable diesel engines were already designed and there was immense experience of track-laying vehicle design and production available within the company.

Michael Lane's excellent history of the Fowler company, *The Story of the Steam Plough Works*, states that the plans had the agreement of the Ministry of Agriculture and the Board of Trade and the production of twelve prototype tractors was authorised.

This was the plan which Rotary Hoes took over. Four tractor models were involved, covering the market from 12 hp to 54 hp. The FD2, with 24 hp, was expected to be the most important model in the range and was allocated 50 per cent of the planned production volume.

The power unit was the Fowler DU four-cylinder engine. It was started by hand, using an inertia device and a decompressor. A range of six gears forward and reverse provided travel speeds up to 4.6 mph forwards and 1.37 mph in reverse, at the governed maximum engine speed of 1250 rpm. The maximum speed in the bottom ratio reverse gear was 0.05 mph. There was a controlled differential steering mechanism operated by two hand levers.

An unusual feature of the FD2 was a power take-off from the side of the transmission housing. This was a rather clumsy attempt to provide a drive shaft to the rear of the tractor in order to operate a 42 in rotary cultivator.

The side drive appears to have been added after the Rotary Hoes takeover and suggests that the FD2 was regarded, to some extent at least, as an aid to launching the idea of rotary cultivation. When the tractor was announced, the principles of rotary cultivation were strongly emphasised and advertising for the tractor continued the same theme. This close association, which is not particularly logical in agricultural terms, may not have been the most positive way to encourage interest in the tractor.

In January 1946, less than a year after the takeover, Rotary Hoes sold the Fowler company to the Thomas Ward group, where it was merged with Marshall of Gainsborough. Marshall, a traditional competitor of Fowler, already had a successful tractor on the market with a single-cylinder diesel engine. In spite of blandly encouraging statements at the time of the merger, this was the beginning of the end for Fowler. The name was used for several years, but the FD tractor range was quickly abandoned and the replacement was the VF tracklayer, with a Marshall engine and Marshall styling.

Under different circumstances, the FD2 might have been a success story, helping to re-establish Fowler as a leading tractor company. The Fowler name still carried immense prestige in some important markets and there was a strong post-war demand for tractor power, including tracklayers. The FD2 seems to have been well designed and the engine was highly regarded.

Instead, it achieved the dubious distinction, to many people, of being the last genuine Fowler product as the company moved towards its final stage of decline.

KENDALL

There were many people in Britain who believed a big market was waiting for a really cheap, lightweight tractor. One of them was Denis Kendall, who announced ambitious plans in March 1945.

He had designed a three-wheel tractor and a light car, both to be powered by an unconventional engine. The car, tractor and engine were all to be built under the Kendall-Beaumont name in a factory at Grantham, Lincolnshire.

His initial announcement included brief details of the engine, which was to be a three-cylinder unit with the cylinders in a radial design around a central crankshaft. The power output was planned as 6 hp.

The Kendall three-wheeler.

The advantages claimed for the three-wheel design of the tractor were low cost and the ability to turn in its own length. Production was planned to reach 2500 tractors a year in the first phase of the programme.

A further announcement came in May, when production was planned to start within 3 months. By this time, the decision had been taken to raise the engine power to 7 hp and further details of the power unit were released. The cylinders would be arranged at a 120° angle, with a total capacity of 595 cc. The design included poppet valves and a hemispherical cylinder head. A supercharged version would be available as an option, Mr Kendall promised.

About 12 months later, there was further news about the tractor. It was still not in full commer-

cial production, but a small number had been produced. The engine was now an air-cooled Douglas twin cylinder, rated at 8 hp and producing a claimed 1400 lb of drawbar pull on a hard surface. The specification included three forward ratios and a reverse, with a maximum speed of 7 mph.

The tractor was now called a Kendall and the manufacturing company was named Grantham Productions Ltd. A retail price had still to be fixed, but £180 was forecast as the likely figure.

Production continued on a small scale through the second half of 1946, while the manufacturers were struggling with financial problems. The end came in 1947, when Grantham Productions went into liquidation. The factory and other assets were bought by Newman Industries of Bristol, who re-equipped with the 'most modern machinery' and produced a range of small three- and four-wheel tractors for several years.

The Kendall car never arrived on the market and the Kendall radial engine apparently progressed no further than the prototype stage. At least four of the tractors have survived and these are currently preserved in museums or collections.

BRISTOL 20

The best years for tracklayers in Britain came after the end of the war in 1945, when tractors of all types were selling in large numbers, and before the arrival of improved four-wheel drive models in the mid-1950s.

This was the period in which sales of Bristol

A Bristol 20 tractor in a photograph in the archives of the National Institute of Agricultural Engineering at Silsoe, Bedfordshire.

tractors reached their peak. The company had started building crawler tractors during the early 1930s, catering for market gardens and small-holdings with a mini-tractor powered by an air-cooled 10 hp engine. This model, and various derivatives of it, sold in modest numbers until World War 2 brought a temporary end to production.

After the war, the Bristol company was operating under new ownership with a change of policy. The new owner was the H. A. Saunders group, a leading distributor of Austin cars and trucks, and their major policy change was to build a more powerful tractor which would have a wider potential market.

A result of the new policy was the Bristol 20, powered by a modified, four-cylinder Austin car engine, which developed just over 20 hp. The new tractor was equipped with power take-off shaft and hydraulic linkage was available as an optional extra at £47.10s (£47.50p). Another option was an electric start kit, which added £22.10s (£22.50p) to the tractor's basic price of £480 when it was announced in 1947.

One feature retained from earlier Bristol models was the use of rubber-jointed tracks. These were made up with castings separated by compressed-rubber blocks. One disadvantage with this arrangement was that the tracks were liable to be shed on some irregular surfaces and refitting them was a complicated procedure.

There were three forward gear ratios, with the drive from the gearbox divided through two clutches. Each clutch transmitted power through a set of reduction gears to one of the front track sprockets. To steer the tractor, the appropriate clutch was disengaged, allowing the other track to continue driving the tractor. There were also steering brakes for use when sharper turns were needed.

A new model, the Bristol 22, replaced the 20 in 1952. This brought increased power, improved track design and variable track width; a Perkins P3 diesel engine was available as an option from 1953 onwards. By this time, the rubber-jointed tracks had been redesigned to reduce the risk of shedding and to make it easier to refit a track which did slip off the sprockets.

Sales of crawler tractors, including the Bristol range, lost ground from the mid-1950s and have never recovered the peak levels of the period when the Bristol 20 was on the market.

HOWARD PLATYPUS

Following their brief adventure with Fowler of Leeds, the Rotary Hoes directors formed a new subsidiary company to build crawler tractors in a factory in Basildon, Essex. The subsidiary was the Platypus Tractor Co., which began production in about 1950.

The design for the new tractors was based on the ideas of Mr. A. C. Howard, who had developed the Rotavators which were the principle product of the Rotary Hoes group. Mr. Howard had been born in Australia and he may have suggested using the name 'Platypus', which is a small animal found in Australia.

The manufacturers explained the choice of name in the introduction to the instruction book which reads as follows:

'Small yet very powerful for his size, the Platypus is a tireless worker. He is as much at home in the water as on land. His webbed feet can dig and handle soil as well as any human-made tool. Add that Platypus comes from two Greek words meaning "flat-footed" and you can easily understand why this name was chosen for our crawler tractor.'

Platypus tractors were designed to take various Perkins diesel engines and a Standard petrol engine was also available as an alternative. The top-selling model was the Platypus 30, which was powered by a P4 developing slightly over 30 hp.

(Overleaf) The Bogmaster version of the Howard Platypus with extra-wide tracks.

Publicity for the Platypus emphasised using the tractor with a Rotavator and a power take-off was standard equipment. But the Platypus was obviously a general-purpose tractor, which was suitable for a wide range of agricultural and civil engineering work.

One of the most interesting versions of the Platypus was the Bogmaster, designed to work in conditions which were too soft for a conventional tractor. The Bogmaster tracks were lengthened towards the rear and formed of extra-wide 32-in track plates. The makers claimed a ground pressure of less than 1.3 psi under the Bogmaster tracks from a tractor weight of 2.5 tons.

A powered trailer was designed to work with the Bogmaster. This used the same track size as the tractor, with a drive shaft from the power

(Right) The standard model Platypus with narrow tracks, powered by a Perkins P4 diesel engine.

(Below) This was an experimental version of the Platypus built at the Basildon factory for trials with a Perkins L4 engine.

The wide-tracked Bogmaster version of the Platypus photographed with the special trailer fitted with tracks powered from the tractor power take-off shaft.

take-off to a crownwheel on the rear axle of the trailer.

The Bogmaster and some of the powered trailers were exported in encouraging numbers, with the Irish peat industry as a substantial customer. The demand encouraged Mr. Howard to design a new self-propelled transport vehicle based on the Bogmaster. This had the driver's seat moved to the front of the vehicle, with a large load-carrying platform at the rear. The trans-porter was called the Bogwaggon.

A 71 hp version of the Platypus, the PD4, was added to the product range, powered by a Perkins R6 diesel, and there was a prototype 50-hp model with an L4 engine.

In spite of the ambitious and imaginative development programme, the Platypus project was failing to meet the company's objectives. There were some teething problems, especially in the track design, which were expensive to rectify, and the crawler-tractor market was becoming increasingly competitive.

Rotary Hoes decided to close the Platypus operation in about 1958 to concentrate on the rapidly growing market for Rotavators.

ALLIS–CHALMERS FUEL CELL TRACTOR

Fuel cells are an old idea which attracts renewed interest from time to time. The basic theory was described by a British scientist in 1839, but was regarded as little more than a curiosity with little practical value for more than a century.

A fuel cell converts chemical energy into electric power. It consists of electrodes and an electrolyte and is supplied with fuel, which may be in the form of a mixture of gases, a solid material or a liquid. Unlike other types of battery cell, the fuel cell has no storage capacity. The production of electrical energy takes place in-stantaneously, as the fuel is supplied, and the electricity has to be used as it is produced.

Development work has been encouraged because fuel cells have a very high theoretical efficiency. When diesel fuel is burned in an engine, 60 per cent or more of the energy in the fuel is likely to be wasted, mostly in the form of heat losses. Energy losses could be 10 per cent or less when a fuel is used in a fuel cell.

Allis Chambers built this tractor to demonstrate their fuel cell power unit.

The driver operates the control lever which governs the forward speed of the tractor and makes the electric motor run forwards or in reverse.

Allis–Chalmers was one of several companies investigating fuel-cell development 20 years or more ago. One of their research projects involved the use of fuel cells to power farm tractors. A standard D-12 tractor was converted to fuel-cell operation and in 1959 the tractor illustrated was built especially for the fuel-cell programme.

The fuel supplied to the cells was a mixture of gases, with propane as the principle ingredient. The gases were carried in cylinders, stowed beneath the rear axle and behind the driver's seat, and were fed through tubes to the cells which were in the space normally occupied by an engine.

In the experimental tractor, there were 1008 individual fuel cells, linked up in groups of nine and arranged in four main banks. Inside the cells, the gases reacted in an electrolyte, helped by a catalyst, with a direct current produced as the reaction took place. The current was fed through an external circuit to a 20 hp direct-current electric motor, which provided the power to operate the tractor for jobs such as ploughing.

Publicity material prepared for the original press demonstration claimed that the tractor was easy to operate. A single lever controlled the amount of current fed to the motor by placing the four banks of cells in series or in parallel. This governed the forward speed and avoided the complication of a gearbox. A second lever allowed the driver to select forward or reverse travel simply by changing the polarity of the current to the motor.

In spite of the theoretical gain in fuel efficiency, the fuel-cell tractor is now a museum piece at the Smithsonian Institution in Washington DC. Problems of cost, poor power to weight characteristics, and some difficulties over the performance of the cells have helped to maintain the supremacy of the diesel engine on the farm.

INTERNATIONAL HARVESTER HT–340 GAS TURBINE TRACTOR

This tractor was built partly for engineering research purposes and partly to attract publicity. Having achieved both of these objectives, International Harvester sent the tractor to the Smithsonian Institution, Washington DC, where it is part of the agricultural collection.

The most noticeable feature of the tractor is the streamlined shape. The styling was designed for publicity purposes, like the driver's crash helmet. Much more interesting is the fact that the bodywork was moulded in fibreglass, rustproof and permanently coloured. Underneath the eye-catching exterior, the engine and transmission were unique. The main purpose of the tractor was to investigate the use of a gas turbine power unit for agricultural purposes and this was linked with a hydrostatic transmission.

The tractor was first demonstrated in 1961. At that time, gas turbines were being confidently forecast as the power unit of the future. Turbo-prop airliners were setting completely new standards of speed and smoothness and the Rover company in Britain had scooped massive publicity with the first turbine-powered car.

Most of the large car companies in Europe and the USA quickly followed Rover's example with a series of turbine cars. Problems of noise, fuel consumption and high manufacturing cost were brushed aside and the smoothness, reliability and favourable power-to-weight characteristics of the turbine were stressed instead.

Several companies experimented with turbine power for farm tractors, but International took the lead and gained the most publicity. They

International Harvester's experimental gas-turbine tractor.

were helped by the fact that an IH subsidiary was already in the gas turbine business and had developed a small turbine for use in a helicopter. This was the engine which was used in the tractor research programme.

The turbine developed 80 hp, although it was rated down to 40 hp to suit the limitations of the HT-340 transmission. The power unit weighed only 90 lb, including the reduction gearing, with an overall length of 21 in and a 13 in diameter.

International had developed the hydrostatic transmission to operate with a piston engine of 40 hp, but claimed it was a particularly suitable type of transmission to work with a gas turbine. The turbine powered a variable displacement pump which produced an oil flow to radial hydraulic motors built into each driving wheel.

The turbine was designed to operate at a constant speed, which is the efficient way to use a turbine. This considerably simplified the controls, which consisted of an electric start button, a steering wheel, brakes and the transmission lever, which controlled the oil flow – and thus the speed – and also selected forward or reverse travel.

A press handout issued by the company emphasised the very considerable attractions of the turbine, which included a long service interval and the excellent forward vision which could be achieved with such a compact unit. But they were also refreshingly frank about the fuel-consumption problems and the loud engine noise.

These and other problems have so far kept the gas turbine out of the tractor market, in spite of some potentially attractive characteristics. Now there is little enthusiasm for the idea of turbine-powered tractors and the diesel engine appears to be secure as the power unit for the future.

COUNTY SEA HORSE

If you happen to need a tractor which floats, speak to County. They are one of very few companies with practical experience of building a tractor which is completely amphibious.

Several companies have produced special tractors with a high degree of waterproofing to work in extremely wet conditions. One example is the Water Buffalo, designed in Scotland by Cuthbertson in 1951 and built in small numbers for land reclamation in marshland. A watertight hull was claimed to allow the Water Buffalo to work in up to 4 ft of water.

County took the matter much further when they designed the Sea Horse. It was based on a County four-wheel drive derivative of the Fordson Super Major, with a 52 hp Ford engine and Ford gearbox.

The Sea Horse version was produced in 1963–4. The flotation was achieved mainly by fitting oversize tyres and making a sealed compartment in each wheel. The tyres were 23.1/18–26 Good Year TD-7s. The tread pattern, which produced the wheelgrip to pull a plough on dry land, propelled the tractor along when it was floating. When the top gear ratio was being used in deep water, the tread pattern produced a substantial amount of splashing and the absence of a cab made the driver's job a very wet one.

Although the tyres and wheel compartments

(Overleaf) The Sea Horse on a trip among the pleasure boats at Fort Lauderdale, Florida.

(Below) The County Sea Horse photographed on its way across the sea from France to England.

County's Sea Horse working in a demonstration on dry land.

were adequate to keep the tractor floating, another problem arose when the Sea Horse was travelling across deep water. This was a serious loss of stability when the tractor met the wash from large ships. It was a problem which most tractor manufacturers had previously ignored and County solved it by adding large sealed compartments at the front and rear ends of the tractor. When the tractor was working on land, these extra compartments could be used for holding extra ballast weight.

Other modifications to make the tractor seaworthy included special seals on the clutch housing, screw-fitting dipsticks, and breathers above the waterline on all of the transmission housings.

County made considerable efforts to publicise the Sea Horse in order to encourage sales. One was shipped to the USSR for demonstration purposes, while another worked in the USA in a tour which included a trip among the pleasure craft at Fort Lauderdale, Florida.

The most ambitious publicity exploit was a voyage across the English Channel from France to the English coast. The crossing took place on 30 July 1963. The 28 nautical miles from Cap Gris Nez on the French coast to Kingsdown took 7 hours and 50 minutes, giving an average speed of approximately 3.6 knots.

KUBOTA TALENT 25

This was another attempt to design a tractor of the future and it included some interesting ideas.

When Talent 25 was first publicised in 1970, Japanese tractors were still a novelty in most western countries and Kubota's futuristic experiment in design attracted little interest.

One of the Talent 25 features which suggested a future trend was the use of low ground-pressue tyres to minimise soil compaction. The level of comfort inside the cab was also ahead of its time, with full air conditioning, a stereo system and heater.

The driver was also provided with a television monitor inside the cab, which was built-in and could be switched through to each of the three cameras built in to the tractor. One camera was at the rear of the tractor, so that the driver could check the operation of trailed equipment. The other cameras were positioned to give a picture from each side of the tractor.

Kubota may have included the closed-circuit television equipment to attract publicity rather than for its practical value. But this type of equipment is now beginning to appear on some machines, such as large harvesters and slurry injection systems, where the driver's vision is obstructed.

Talent 25 was designed for two-way operation, which is an arrangement which Kubota has recently introduced as an option on some of its production models. The driver's seat on the 1970 version was designed to swivel so that the operator faced the right way for working the

Kubota's idea of a tractor for the future, complete with closed-circuit television equipment.

tractor in reverse. The hydrostatic transmission provided the same stepless speed variation forwards and backwards.

One of the less logical design features was the rear access to the cab. This created problems with rear-mounted equipment in place and the double rear doors produced a thick vertical line to obscure the driver's visibility when operating the tractor in reverse.

The power unit for the experimental tractor was a twin-cylinder diesel, horizontal and with air cooling. The output was 25 hp from 1200 cc capacity. This is a type of engine still used on some Kubota industrial equipment.

Steering was by means of two levers, one on each side of the cab, and these remained accessible whichever way the seat faced.

One design feature which still appears to be for the far distant future, was the facility for remote control by means of a radio transmitter. There are few jobs in farming that many of us would trust to a driverless tractor.

Picture Credits

Black and White

Alberta Museum & Archives: p. 69; Allis-Chalmers: pp. 71, 72, 133, 134; County Tractors: pp. 137, 138, 140; Deere & Co: pp. 89, 90, 119, 120; Terence J. Fowler: p. 18 (lower); Greenfield Village & Henry Ford Muesum: p. 93; Howard Rotavator Co: pp. 123, 124, 128, 130, 131, 132; International Harvester: pp. 79, 80, 83, 136; Klöckner-Humboldt-Deutz: pp. 19, 20, 21; Kubota: p. 141; Massey-Ferguson: pp. 111, 112, 114, 117; Andrew Morland: pp. 7, 8, 9, 30, 45, 52, 53, 57, 59, 75, 78, 86, 95, 105, 115, 125; Museum of English Rural Life: pp. 13, 17, 22, 34, 35, 37, 41, 43, 47, 56, 63; National Institute of Agricultural Engineering: p. 126; Renault: p. 108; SAME: pp. 102, 103, 104; E.J. Tippett: pp. 96, 97, 98, 99, 100, 101; Volvo BM: pp. 32, 33; Western Development Museum: p. 18 (upper); David Williams: pp. 121, 122.

Colour

Terence J. Fowler: Plates 2, 3, 7, 8, 14, 16, 17, 20, 22, 29, 31; Andrew Morland: Plates 1, 4, 9, 10, 11, 12, 13, 15, 18, 19, 21, 23, 24, 26, 27, 28, 30; Ontario Agricultural Museum: Plate 6; SAME: Plate 25; Volvo BM: Plate 5.

Other pictures are from the author's collection.

INDEX

Figures in **bold** refer to colour plates and those in *italic* to black and white illustrations.